Inspection, Evaluation and Maintenance of Suspension Bridges
Case Studies

Inspection, Evaluation and Maintenance of Suspension Bridges
Case Studies

Edited by

Sreenivas Alampalli
New York State Department of Transportation

William J. Moreau
Formerly Chief Engineer, New York State Bridge Authority

CRC Press
Taylor & Francis Group
Boca Raton London New York

CRC Press is an imprint of the
Taylor & Francis Group, an **informa** business
A SPON PRESS BOOK

CRC Press
Taylor & Francis Group
6000 Broken Sound Parkway NW, Suite 300
Boca Raton, FL 33487-2742

First issued in paperback 2019

© 2016 by Taylor & Francis Group, LLC
CRC Press is an imprint of Taylor & Francis Group, an Informa business

No claim to original U.S. Government works

ISBN-13: 978-1-4665-9688-7 (hbk)
ISBN-13: 978-0-367-86853-6 (pbk)

Visit the Taylor & Francis Web site at
http://www.taylorandfrancis.com

and the CRC Press Web site at
http://www.crcpress.com

To all the bridge engineers around the world
for their dedicated service to public safety.

Sreenivas Alampalli and William J. Moreau

Contents

Preface

Owners and operators of long-span suspension bridges are scattered around the globe, just like the bridges they maintain. Some of the oldest long spans are in the United States; the newest are in and around the Pacific Rim; and Europe is home to many world-class suspension bridges, with many built post World War II. New construction continues throughout Europe in the Netherlands, the United Kingdom, and Turkey. An array of interesting concepts to connect Italy with the island of Sicily is also on the drawing board.

A challenge for long-span suspension bridge operators is that most operators oversee only one or two bridges through small single-purpose public authorities or subdivisions of state-run transportation agencies. This, combined with the fact that most suspension bridges operate over a very long lifetime, makes it difficult for bridge owners to acquire the experience necessary to detect signs of deterioration early, develop effective mitigation plans, and implement the appropriate restoration in a timely and cost-effective manner.

The International Cable Supported Bridge Operators' Association was conceived in 1991 when over 125 international suspension bridge owners and operators assembled in Poughkeepsie, New York, to discuss common concerns, present research papers, and observe the main cables of the Mid-Hudson Bridge, which was undergoing a full-length main cable rehabilitation project at the time. Attendees traveled from Europe, South America, Asia, and across the United States to share problems, solutions, and best practices with the goal of reducing this challenge.

The companion volume, which was published in 2015, assembled decades of knowledge and experience through the authorship of many progressive suspension bridge owners. Based upon their own perspectives, each owner discusses their state-of-the-practice for suspension bridge engineering, including the nuances of each bridge element unique to suspension bridges, together with a historical overview, design, inspection, evaluation, maintenance, and rehabilitation.

This volume illustrates historical to current operations of selective suspension bridges all around the world in detail as told by an outstanding array of international bridge operators. The resurgence of the suspension bridge as a practical bridge type has been brought about through a confluence of changes. These changes consist of new materials, new construction methods, and the desire to cross geography with unsuitable foundation locations or depths. Manufacturing centers and shipping routes have also changed considerably throughout the world, thereby making the cost advantages of suspension bridges economically viable once again.

While the number of suspension bridges opened to traffic in the second half of the 20th century was relatively small, the turn of the century has seen dozens of new spans open, some with remarkable span lengths. This series of case studies covers the generic day-to-day issues of suspension bridge inspection and routine maintenance, as well as periodic inspection, maintenance, and evaluation that may uncover some hidden concerns. Timeliness is crucial in arresting the various degradation processes that begin to attack the vulnerabilities of suspension bridges from the day they are built. Careful documentation of the conditions found during these inspections will be invaluable and will be intently studied by future bridge tenders. Trial and error has taught us that many layers of paint on the cable covering may not be the best solution to cable protection. Misconceptions as well as success stories are shared in the hope of advancing the state of the practice for bridge owners and operators.

The principal objective for all of us in transportation is to protect the public safety while enhancing the mobility of our communities for economic and quality-of-life improvements. However, when working with such demonstrable examples of humanity's engineering abilities, we tend to want to ensure a perpetual service life for the grandest engineering icons. Lessons learned over time will be our best asset in improving our performance in this regard. We thank all who were involved in the authorship and development of this book. Their personal efforts and contributions to the industry will not be forgotten.

Sreenivas Alampalli and William J. Moreau
Editors

Acknowledgments

It has been a great pleasure working with owners of suspension bridges around the world to bring this book to fruition. The discussions for documenting the body of knowledge gained by the current and past suspension bridge owners originated at a meeting of the International Cable Supported Bridge Operators Association workshop held in May 2012 in New York State. It was decided by the editors that documenting these experiences in the form of two books would serve not only the owners but also the entire bridge industry. The companion volume discusses the state of the practice in suspension bridge inspection, evaluation, and rehabilitation methods used worldwide. This volume discusses specific bridges around the world to give a comprehensive picture of how these suspension bridges are operated and maintained by an array of countries and cultures. Knowing that the contributing authors did not have time to undertake such an effort as this but spent valuable time in documenting their experiences and practices for future generation of engineers, we thank them for their time and efforts in making this book, and the companion book that was published earlier, a reality.

We also thank our families (Sharada Alampalli and Sandeep Alampalli and Cheryl Moreau) for their support during the preparation of these books.

Sreenivas Alampalli and William J. Moreau

Editors

Dr. Sreenivas Alampalli, PE, MBA, is director of the Structures Evaluation Services Bureau of the New York State Department of Transportation. Before taking up the current responsibility in 2003, Dr. Alampalli was director of the Transportation Research and Development Bureau, where he worked for about 14 years in various positions. He also taught at Union College and Rensselaer Polytechnic Institute as an adjunct faculty member.

Dr. Alampalli obtained his PhD and MBA degrees from Rensselaer Polytechnic Institute; his MS degree from the Indian Institute of Technology, Kharagpur, India; and his BS degree from Sri Venkateswara University, Tirupati, India. His interests include infrastructure management, innovative materials for infrastructure applications, nondestructive testing, structural health monitoring, and long-term bridge performance. Dr. Alampalli is a fellow of the American Society of Civil Engineers, the American Society for Nondestructive Testing, and the International Society for Health Monitoring of Intelligent Infrastructure. He is the recipient of the Bridge Nondestructive Testing Lifetime Service Award in 2014 from the American Society for Nondestructive Testing for outstanding voluntary service to the bridge and highway nondestructive testing industry. In 2013, he also received the American Society of Civil Engineers' Henry L. Michel Award for Industry Advancement of Research. Other notable awards he received include the American Society of Civil Engineers' Government Civil Engineer of the Year in 2014, and the prestigious Charles Pankow Award for Innovation from the Civil Engineering Research Foundation in 2000. He has authored or coauthored more than 250 technical publications, including two books on infrastructure health in civil engineering.

Dr. Alampalli is an active member of several technical committees in the Transportation Research Board, the American Society of Civil Engineers, and the American Society for Nondestructive Testing. He currently chairs the Technical Committee of Transportation Research Board on field-testing and nondestructive evaluation of transportation structures. He served as the Transportation Research Board representative for the New York State Department of Transportation and as a member of the National Research Advisory Committee. He is a book review editor of the American Society of Civil Engineers' *Journal of Bridge Engineering* and serves on the editorial boards of the journal *Structure and Infrastructure Engineering: Maintenance, Management, Life-Cycle Design and Performance* and the journal *Bridge Structures: Assessment, Design and Construction.*

William J. Moreau, PE, served as the chief engineer of the New York State Bridge Authority for over 27 years. Two of the six Hudson River crossings previously under his care were suspension bridges, constructed circa 1924 and 1930. Maintenance and preservation of these world-class suspension bridges became a career objective for Moreau, culminating with service on the peer-review panel for the development of National Cooperative Highway Research Program Report 534, *Inspection and Strength Evaluation of Suspension Bridge Parallel-Wire Cables.* Moreau participated for many years as a member of Transportation Research Board committees and served as a chairman of the Construction Committee for Bridges and Structures from 2005 through 2007.

Many early main cable-inspection techniques, evaluation methods, and restoration materials were developed through partnerships Moreau developed with truly outstanding members of the engineering, construction, and material manufacturing communities in pursuit of arresting the effects of age and environment on the main cables of the Hudson Valley suspension bridges. He is currently semiretired and continues consulting in the New York City area.

Contributors

Sreenivas Alampalli
New York State Department of
  Transportation
Albany, New York

Finn Bormlund
A/S Storebælt
Copenhagen, Denmark

Anna Chatzifoti
Halifax Harbour Bridges
Dartmouth, Nova Scotia, Canada

Ahsan Chowdhury
Halifax Harbour Bridges
Dartmouth, Nova Scotia, Canada

Barry Colford
AECOM
Philadelphia, Pennsylvania

Jon Eppell
Halifax Harbour Bridges
Dartmouth, Nova Scotia, Canada

James D. Gibson
TIML MOM Ltd
Hong Kong, China

Brian Gill
New York City Department of
  Transportation
New York, New York

Mary Hedge
MTA Bridges and Tunnels
Metropolitan Transportation
  Authority
New York, New York

Justine Lorentzson
MTA Bridges and Tunnels
Metropolitan Transportation
  Authority
New York, New York

William J. Moreau
HAKS, Engineering
New York, New York

Kim Agersø Nielsen
A/S Storebælt
Copenhagen, Denmark

Katsuya Ogihara
Honshu-Shikoku Bridge
  Expressway Co., Ltd.
Sakaide, Japan

Dora Paskova
MTA Bridges and Tunnels
Metropolitan Transportation
  Authority
New York, New York

Mohammad Qasim
MTA Bridges and Tunnels
Metropolitan Transportation
    Authority
New York, New York

Stewart Sloan
Port Authority of New York/
    New Jersey
New York, New York

Leif Vincentsen
A/S Storebælt
Copenhagen, Denmark

Judson Wible
Parsons Brinckerhoff, Inc.
New York, New York

Gongyi Xu
China Railway Major Bridge
    Reconnaissance and Design
    Institute (BRDI)
Wuhan, China

Bojidar Yanev
New York City Department of
    Transportation
New York, New York

Jeremy (Zhichao) Zhang
MTA Bridges and Tunnels
Metropolitan Transportation
    Authority
New York, New York

Chapter 1

Manhattan Bridge

Bojidar Yanev and Brian Gill

CONTENTS

1.1 DESIGN AND CONSTRUCTION

After the Brooklyn (1886) and Williamsburg (1903) Bridges, the Manhattan was the third East River suspension bridge to provide vehicular and rail traffic between the New York City boroughs of Brooklyn and Manhattan. It was opened officially on December 31, 1909, by Mayor George B. McClellan, Jr., whose term was expiring on that date. About 30 m (100 ft) of the bridge lower roadway over Division Street in Manhattan consisted of temporary planking to allow the passage of the mayor's motorcade

(*New York Times*, January 1, 1910). The Second Avenue elevated portion of the subway had to be lowered 6 ft over a length of 244 m (800 ft) to accommodate the bridge clearance (*New York Times*, December 5, 1909) in that area.

The Manhattan Bridge is 1761.4 m (5779 ft) long between abutments at the lower level and 1855 m (6086 ft) between portals on the upper levels. Both approaches are supported by three- and four-span continuous Warren trusses. Several stringer and floor beam spans support the upper roadways between portals and abutments. The main suspension bridge is 890 m (2920 ft) long, with a main span of 448 m (1470 ft) and two 221 m (725 ft) side spans. Four 7.3 m (24 ft) deep stiffening trusses (designated as A, B, C, and D from south to north) run between abutments. These are supported by piers on the approaches and by the four main cables on the suspended spans. Their spacing is 8.5 m–12.2 m–8.5 m (28 ft–40 ft–28 ft). The Brooklyn and Manhattan bound upper levels rest on trusses A–B and C–D, respectively. All other traffic is at the lower chord level. Figure 1.1a shows the original elevation and cross section of the bridge along with some details related to its construction. Figure 1.1b illustrates its location across the East River relative to the Brooklyn Bridge downstream.

As illustrated in Figure 1.2, the bridge has always carried the most people of any East River crossing. Originally, it was designed for railroad on the upper level, trolley cars underneath, and vehicular traffic on a wood-block deck in the center of the lower level. The structure now supports four vehicular lanes on the upper level, three lanes of vehicular traffic, four subway transit tracks, and a bikeway and a walkway on the lower level. Recent traffic counts surpass 500,000 commuters on weekdays (110,000 passengers in 85,000 vehicles, 390,000 mass transit riders, and 6000 bikers and pedestrians). Figure 1.3a and b shows general views of the bridge.

1.1.1 The transportation demand

The need for an all-railroad bridge was first suggested in the summer of 1895 by James Howell, former New York City mayor and later president of the Brooklyn Bridge Board of Trustees, as a measure to relieve congestion on the Brooklyn Bridge (Nichols, 1906). At the time, rail travel had much more influence on public policy than vehicular travel had. Manhattan Bridge would be the first railroad bridge to connect Long Island, the most populated island in the United States, with the mainland in a combination with a Hudson River crossing. The latter would be Gustav Lindenthal's 869.25 m (2850 ft) long suspended braced-eyebar bridge carrying several railroad tracks crossing the Hudson River first at Canal Street, then at 10th Street.

John Mooney, Secretary for the Board of Public Improvements noted (New York City Department of Bridges, 1904, pp. 341–342), "By removal of comparatively few buildings of poor quality and low cost, the solving of the problem of a straight line thoroughfare from the junction of Atlantic

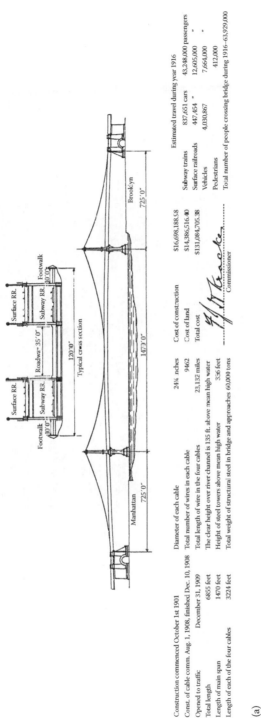

Construction commenced October 1st 1901
Const. of cable comm. Aug. 1, 1908, finished Dec. 10, 1908
Opened to traffic   December 31, 1909
Total length   6855 feet
Length of main span   1470 feet
Length of each of the four cables   3224 feet

Diameter of each cable   24¾ inches
Total number of wires in each cable   9462
Total length of wire in the four cables   23,132 miles
The clear height over river channel is 135 ft. above mean high water
Height of steel towers above mean high water   336 feet
Total weight of structural steel in bridge and approaches 60,000 tons

Cost of construction $16,698,188.58
Cost of land $14,386,516.90
Total cost $131,084,705.38

signature
Commissioner

Estimated travel during year 1916

Subway trains	837,651 cars	43,248,000 passengers
Surface railroads	447,454 "	12,605,000 "
Vehicles	4,030,867	7,664,000 "
Pedestrians		412,000
Total number of people crossing bridge during 1916–63,929,000		

Typical cross section

Footwalk Subway RR. | Roadway 35'0" | Subway RR. | Footwalk
Surface RR. | | Surface RR.
120'0"
10'0"

Manhattan
725'0"   147'0"   725'0"   Brooklyn

(a)

Figure 1.1 (a) Manhattan Bridge, 1909.

(*Continued*)

Figure 1.1 (Continued) (b) Brooklyn and Manhattan Bridges across East River.

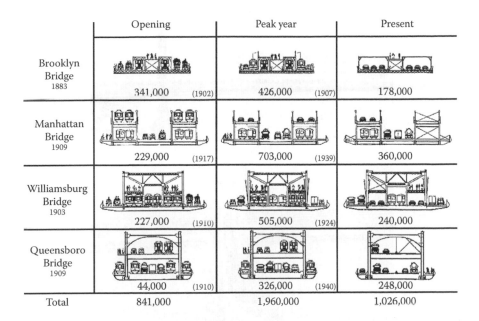

	Opening		Peak year		Present
Brooklyn Bridge 1883	341,000	(1902)	426,000	(1907)	178,000
Manhattan Bridge 1909	229,000	(1917)	703,000	(1939)	360,000
Williamsburg Bridge 1903	227,000	(1910)	505,000	(1924)	240,000
Queensboro Bridge 1909	44,000	(1910)	326,000	(1940)	248,000
Total	841,000		1,960,000		1,026,000

Figure 1.2 Use of the East River crossings from their opening to 1988.

and Flatbush Avenue and the station of the Long Island Rail Road (LIRR), long contemplated ... and from the end of the bridge at Canal Street ... and thence uptown or to the North (Hudson) River." The rail link never materialized and Long Island would have to wait until 1916 for the completion of Lindenthal's signature Hell Gate arch for its only direct rail link to the mainland.

1.1.2 Preliminary designs

By 1898 there were 15 to 20 alignments plotted and six proposed designs for what was called the third East River bridge. Four of these designs featured cantilevered main bridges and two were suspended wire cable bridges, one with a 55 ft high stiffening truss and the other with a 35 ft high truss (Richard S. Buck's design), evoking debates over the most efficient and aesthetic bridge type for the location and intended purpose.

In addition to the cantilever/suspension debate for the best design of long-span bridges unfolding during this period, another debate was playing out between the use of braced eyebars versus wire cable–supported suspension bridges. This debate, heated at times, resulted in three separate design proposals between 1899 and 1904 and, along with changes to user funding, delayed the construction of the bridge by several years.

In November 1899 Mayor Van Wyck met with the Board of Public Improvements and noted that "after mature deliberation, it was decided

(a)

(b)

Figure 1.3 (a) Manhattan and Williamsburg Bridges across East River and (b) Manhattan Bridge viewed from the Brooklyn Tower, 2012.

to adopt the suspension bridge. The location is so close to the present New York and Brooklyn Bridge that any departure in style or type of structure would not be pleasing or commendable" (New York City Department of Bridges, 1901, p. 336).

The first design proposal was for a wire cable suspension bridge with 10.67 m (35 ft) high stiffening trusses. It was designed by Richard S. Buck, chief engineer in charge of the newly created New York City Department of Bridges, and approved by the Board of Public Improvements in November 1899. The bridge was to be 2813.3 m (9230 ft) long from Canal Street and the Bowery in Manhattan to Willoughby and Price Streets in Brooklyn (New York City Department of Bridges, 1901, p. 266), with a 2.8% maximum grade (4% was the built design). If constructed, it would have eliminated about one-half of the length of the Flatbush Avenue Extension that ended at LIRR's Atlantic Terminal.

Work on the first approved design actually began. The tower foundation contracts were advertised and constructed based on this plan (Nichols,

1906). The tower foundations were later adapted to accommodate the newer design by additional masonry (Johnson, 1910, p. 22).

Richard S. Buck (New York City Department of Bridges, 1901, p. 363) noted, "No attempt has been made to complete plans of any part of the work much ahead of the time they are to be executed. It has been thought best rather to cover as much ground as possible in careful studies of all controlling features of the design in order that all parts of the work may be harmonized as thoroughly as possible."

The second design was advanced by G. Lindenthal after he was appointed commissioner by Mayor Seth Low in 1902. Lindenthal proposed changing the entire character of the bridge to a braced eyebar suspension bridge in March 1902 (Nichols, 1906, p. 23). The capacity of the bridge was also increased by adding two elevated tracks (New York City Department of Bridges, 1904, p. 133). Lindenthal's eyebar design was demonstrably feasible, as the 290 m (951 ft) main span of the Elisabeth Bridge in Budapest was being constructed. In 1903 that was the longest chain-supported span in the world. Saint Mary's, built in 1929, was the last chain bridge built in the United States. The last European chain bridge was built in Cologne in 1915 (Griggis, 2008, p. 277). The longest suspended eyebar span at 340 m (1115 ft) is the Florianópolis Bridge, which was completed in 1926 (currently closed to traffic).

Lindenthal noted that using the eyebars would save months, if not years, in reduced construction time based on previous performance of time needed to spin the wire cables (Reier, 1977, pp. 52–53). The eyebar substitute was approved by the Art Commission in March 1903 (New York City Department of Bridges, 1904, p. 22). The length of each eyebar was about 13.7 m (45 ft), compared with 15.25 to 17.7 m (50 to 58 ft) long bars for the cantilever truss of the Quebec Bridge (Nichols, 1906, p. 40).

Lindenthal's eyebar design may have sought justification in Roebling's perceived slow fabrication and spinning on the Williamsburg Bridge, Roebling's largest bridge contract to date (Winpenny, 2004, p. 85; Zink and Hartman, 1992). More to the point, Lindenthal, as much as Waddell, of whom he was dismissive, demonstrated a lifelong preference for eyebars over cables. Thus, the East River bridges in the 21st century testify to the superiority of Roebling's 19th-century vision over the skill of some top early-20th-century professionals.

To compete more effectively with eyebars, the Roeblings were expanding production of their high-strength wire and had started construction of a new plant that would employ 300 workers. During the spinning of the Williamsburg Bridge cables, the Roeblings were not producing their own steel and had to rely on others for delivery of the billets (Zink and Hartman, 1992). There had been inexperience in working the special steel into the dimensions and length described (Nichols, 1906, p. 27). At the time, there were 11 companies in the United States that could produce nickel steel eyebars, but Roebling & Sons was the only producers of the wire specified (Reier, 1977, p. 53).

Opponents to the eyebar design noted that the Elisabeth Bridge eyebars were cut from plate and not forged as they would have to be for the larger and heavier Manhattan Bridge span, making comparison of the two designs less valid. Calculations show that there are 10,000 tons more steel required for the eyebar design (New York City Department of Bridges, 1904) Richard S. Buck challenged Lindenthal's arguments about the wire cable design costs and about additional construction time requirements (Griggis, 2008).

Upon his appointment as commissioner of the Department of Bridges in 1904, George Best took note of Roebling's increased production capacity (New York City Department of Bridges, 1904, p. 12): "I am convinced that the wire cable suspension bridge can be built in one-half the time, and at very much less the cost, than the eyebar bridge ... and that a wire cable bridge was anticipated in the original authorization."

Commissioner Best also noted (Nichols, 1906, p. 29), "I am well aware that a commission of celebrated engineers passed favorably upon the design for the eyebar chain bridge, and I am far from denying that a structure of that type can be built at this site. However, this commission made no technical comparison between the two types of bridges and their incidental remark that a chain bridge could be built more cheaply than a cable bridge must be regarded as mere expression of personal preference, because there are absolutely no data in existence from which to determine with the remotest degree of accuracy what the cost of the chain bridge will be in either time or money."

Although much has been written about the eyebar/wire rope design debate, resulting bidding controversy, and the politics of selecting the design, time has shown that wire cables are more redundant and their safety factor more reliably calculated during service. The collapse of the nonredundant eyebar chain–supported Silver Bridge over the Ohio River in 1967 closed the debate.

1.1.3 The third and final design

When Lindenthal was replaced as Commissioner, the eyebar design was replaced with a second wire cable design as the latter was more efficient. In a 1904 letter to the City Art Commission, Commissioner Best wrote (Nichols, 1906), "It is well known that steel reaches is greatest strength when drawn into wire (the weight of the eye bars would be twice the wire cable weight yet only about half the strength) and this combined with the uncertainty in the performance of each eyebar due to the inability to test production pieces makes the wire cable design the preferred design for the new Manhattan Bridge."

The calculations for the redesign were performed by Leon S. Moisseiff, who graduated from Columbia University in 1895 and worked as a draftsman under R. S. Buck on both the Queensboro Bridge and the first Manhattan Bridge design. During the third design, Moisseiff worked under R. S. Buck (who was employed again by the Department of Bridges after

George Best was appointed commissioner) and O. F. Nichols (Griggis, 2008, p. 271). Moisseiff later designed the infamous Galloping Gertie—the original Tacoma Narrows Bridge that collapsed 4 months after opening in 1940. Some features of the tower designed by Lindenthal were retained, but the pinned bases and much of the bracing were removed between the center columns (Griggis, 2008, p. 271).

Moisseiff designed the wire suspension bridge in 6 months by using the newly developed deflection theory to reduce steel weight and cost. This was the first application on a bridge, let alone an eccentrically loaded railroad bridge. Prior suspension bridges were designed with elastic theory, emphasizing deeper trusses (Winpenny, 2004, p. xvii).

The deflection theory, or the "more exact theory," is due to Josef Melan (1888). For further reference, see the other chapters in this book. Prior suspension designs had used the elastic theory developed in 1826 or the Rankine theory developed in 1858. A Fourier series treatment of deflection theory was added in 1930 (Steinman and Watson, 1941). David B. Steinman, another Columbia graduate (1908), noted that the values of the bending moments and shears produced by the elastic theory are too high, thus satisfying safety, but not economy, and that the elastic theory is generally sufficient for short spans with deep rigid stiffening systems (Steinman, 1922). Melan theorized that the maximum span of 4694 m (15,400 ft) was obtainable if the bridge carried only its own weight (Steinman, 1913, p. 17).

According to the deflection theory, the work performed by the truss from dead and live loads equals the total internal work expended in stretching the cable and suspenders and in deflecting or bending the stiffening truss throughout the span. The stiffening truss is erected and adjusted at mean temperature so that the dead load does not produce bending in it (Burr, 1913, p. 212). The moving load is distributed into two parts, the much smaller producing deflections in the stiffening truss and the other a uniform pull on the suspenders, producing cable stresses; these stresses are used in the initial equations (Burr, 1913). Unlike the elastic theory, the deflection theory does not assume that the ordinates of the cable curve remain unaltered under live loads and the lever arms of the cable forces are taken into account (Steinman, 1922, p. 248).

The revised wire cable design was submitted and approved by the Art Commission in September 1904 (New York City Department of Bridges, 1904). The Art Commission noted that it did not have adequate guidelines for accepting bridge designs as it would seem they must consider engineering, economic, and aesthetic factors to make a total comment approving one design over the other. Either was acceptable as long as the new bridge adhered to the architectural effects in Lindenthal's design (Reier, 1977, p. 54).

Fabrication for the superstructure steel for the main bridge began in August 1906. One year later, toward the end of the workday on August 29, 1907, the south arm of the cantilevered Quebec Bridge collapsed, sending 83 workers into the Saint Lawrence River, killing 75 (Winpenny, 2004,

p. 90). The company supplying steel for the Quebec Bridge, Phoenix, happened to be the same as the one that was awarded contracts for steel fabrication and erection of the superstructure of the Manhattan Bridge. Phoenix had the contracts to provide the structural steel for the anchorages, towers, and trusses (Winpenny, 2004, p. 16). Memories also held that Phoenix was involved in construction of the Louisville Bridge, which collapsed in December 1893 during high winds, killing 20 (Winpenny, 2004, p. 27).

Although the construction of a suspension bridge is inherently safer than that of a cantilever bridge, there were justifiable calls for precautions, and in response, the Department of Bridges retained Ralph Modjeski to investigate the Moisseiff design. This included investigating the type of the foundations, stresses in the cable and stiffening truss, corrected dead-load values, and conductivity of heat in the main cables. At the time, the maximum theoretical loading for structural steel was 27,226 kg/m (18,300 lb/ft), which was considered as the practical maximum.

In his report, Modjeski (1909) noted that this rare maximum loading would not reach 80% of the elastic limit stress. The towers and floor system are of carbon steel and the trusses are of nickel steel. This was the first use of nickel alloy steel on a major bridge in significant amounts, including for the riveting. Investigation showed that the first slip of the plates detected 650 to 1000 kg/cm^2 (9500 to 14,670 pounds per square inch [psi]) for field-riveted joints (by pneumatic hammer) and 720 to 1260 kg/cm^2 (10,500 to 18,000 psi) for shop-riveted joints (by a pressure machine). Modjeski observed that had these higher values been known, no doubt some allowance would have been made for stress reversals, resulting in a more efficient design. He concluded that "the structure as a whole has been carefully designed, and when complete will be amply strong to carry the heaviest traffic … as well as any reasonable increase in weight of properly regulated traffic it may be called upon [to support] for many years to come."

The original design loads assumed four lines of crowded LIRR cars, four lines of Brooklyn Rapid Transit cars, four vehicular lanes, and two pedestrian walkways. At an average of 2812 kg/cm^2 (40,000 psi), the yield stresses for the fabricated carbon steel used in the towers were 20% higher than specified. The yield stresses for the fabricated nickel steel trusses averaged at 4289 kg/cm^2 (61,000 psi), or 10% higher than specified.

The suspended structure was designed for dead load, including the cables of 37,180 kg/m (25,000 lb/ft) and a working live load of 11,672 kg/m (8000 lb/ft) or congested live load of 2722 kg (6000 lb) (Perry, 1909, p. 51).

The cables stretch 3 ft due to the maximum dead loading of 29,743 kg/m (20,000 lb/ft), which results in a factor of safety of 2. The cables would have to stretch 9 to 10 ft before the elastic limit was reached (Perry, 1909, p. 65). "The maximum stress on the tower and stiffening truss would occur at congested loading and maximum temperature … Snow loading is offset by the lower temperatures … this principle would not apply to cantilevered bridges."

1.1.4 Construction firsts

The Manhattan Bridge was originally referred to as the third East River bridge, but because of the redesigns and rebidding of contracts, it became the fourth East River bridge to be completed. Even though the construction timeline shows 17 years from the beginning of the tower foundations in 1901 to opening for full service on the Brooklyn Rapid Transit lines in 1918, once the steel tower work started, construction of the towers and superstructures set records and the bridge was substantially completed in 3 years, totaling 42,000 tons between anchorages. Many of the modern construction techniques for suspension bridges were developed and used for the first time on the Manhattan Bridge.

The speed of constructing the main bridge was partially attributed to the fact that the steel towers, cables, suspenders, and suspended superstructure were included in one contract, thereby "eliminating multiplicity of plant, friction between contractors and possible consequent litigation with the City" (Johnson, 1910, p. 28) There had been three contracts let for the main bridge steel of the Williamsburg Bridge. The single contract facilitated orderly fabrication and building of the towers, cables, and suspended spans in an overlapping sequence, without intervals of lost time.

1.1.4.1 Caisson construction

The foundation contracts for the Manhattan and Brooklyn Towers were advertised separately. The caisson for the Brooklyn Tower's foundation was floated into place in February 1902 and the cutting edge rested at an average depth of 27.75 m (91 ft) below mean high water (MHW) or about 18.9 m (62 ft) below the river bottom. The material was described such that it required a pick ax to loosen and was a perfectly reliable foundation. A few cases of the bends developed, two of which were fatal (New York City Department of Bridges, 1904, p. 141).

The 23.8×43.9 m^2 (78×144 ft^2) timber caissons were constructed 13.7 m (45 ft) high in Manhattan for the tower foundation and 17 m (56 ft) high for the Brooklyn foundation to accommodate the plans showing an anticipated depth of 24 m (79 ft) below MHW to a bed of gravel in Manhattan and 28.7 m (94 ft) below MHW in Brooklyn (New York City Department of Bridges, 1901, p. 363).

The Manhattan Tower caisson was floated into place July 1903 and the foundation reached "course sand with fine gravel being very firm in character" at -28.2 m (-92.5 ft) in December 1904 (Modjeski, 1909, p. 4). Attempts were made for weeks to force grout into this material, which was useless, and the pressure of up to 3.2 kg/cm^2 (47 psi) caused the death of several men (Johnson, 1910, p. 26). A study of the conditions resulted in the decision to fill the caissons some 6 m (20 ft) above rock (Nichols, 1906).

1.1.4.2 Towers

In contrast to the Brooklyn and Williamsburg Bridges, which combine relatively rigid towers and sliding saddles, the Manhattan was the first to combine fixed saddles and flexible towers, braced only in the transverse and vertical directions. Moisseiff eliminated Lindenthal's pivot at the tower base. Instead of relying on the rollers under the saddles at the towers, which were largely ineffective on previous bridges, the slender towers resist elastically the varying longitudinal forces caused by ambient service conditions. Under maximum loading and temperature, the actual towers can accommodate a movement of 61 cm (2 ft) each way from the tower tops. Under ordinary conditions the movement was estimated at less than 15 cm (6 in), producing stress in the extreme fiber under 7258 kg (16,000 lb) (Perry, 1909).

Previously unseen in bridge design were also the cellular spaces within the tower legs, replacing exposed elements (such as, for example, at the Williamsburg Bridge). This design allowed construction of the tower columns without falsework. An ingenious derrick could advance vertically up each leg after each 62-ton section was installed (Steinman, 1922, p. 337). The derrick had a platform supported by two struts; the tip moment was resisted by a pair of wheels engaging the vertical edges on the tower. When the 62-ton full section had been added, blocks were added to the top and falls attached to the derrick platform, by which it then lifted to the next level. In addition to the two stiff-leg derricks, each tower had two hoisting engines, a power plant with air compressors, 30 pneumatic riveting hammers, six forges, and a workforce of 100 men and six rivet gangs. This system allowed erecting a record 2000 tons of steel at one tower in 16 working days (Steinman, 1922, p. 165).

In order to offset the deformations caused by congested live loads, the towers were pulled 10 cm (4 in) toward the shores when the cables were completed and prior to placing the dead load (Perry, 1909).

1.1.4.3 Cable spinning

With diameters of 54.2 cm (21 1/4 in), the four main cables were the largest in the world when spun and remained so for 17 years. The two 76.5 cm (30 in) diameter cables on the Benjamin Franklin Bridge were completed by 1926, but only the four cables of the George Washington Bridge, completed in 1931, had greater carrying capacity. At 105 years, the Manhattan Bridge still carries the most traffic and has the largest capacity of all six East River suspension bridges.

Roebling & Sons made good on their marketing promise that the wires for the largest cables in the world would also be spun in record time. In the spring of 1908, the contractor was claiming that the cables would be completed within 12 months of stringing the first wire and at "far greater celerity than the Brooklyn and Williamsburg Bridges." Stringing would be done by late spring 1909 and the bridge would be ready to open by the summer of 1911 (*Scientific American*, 1908).

All cable work was performed by Glyndon Contracting (Perry, 1909). Preparations began with four reels of 4.45 cm (1 3/4 in) wire ropes for the footbridge cables towed across the East River by barge with other traffic stopped. The free end of each rope was hauled up by line over the tower tops, placed on temporary saddles, and adjusted with hoist engines at the Brooklyn anchorage (Perry, 1909, p. 55).

The inner and outer cables were braced by working platforms, and hauling rope towers were stationed every 76.25 m (250 ft). The work platform was stayed against wind vibration, with four 4.46 cm (1 3/4 in) storm cables connected to the footbridge at 16.8 m (55 ft) intervals (Perry, 1909, p. 56).

Guide wires were adjusted to the designed deflection and slippage in the tower and anchorage saddles prior to loading (Hool and Kinne, 1943, p. 350). The four hauling wire ropes featured 1.9 cm (3/4 in) diameter endless loops with two traveling sheaves. The hauling rope at the Manhattan anchorage passed around two 0.915 m (3 ft) diameter deflecting wheels and one 1.525 m (5 ft) diameter idler wheel that could adjust the tension (Perry, 1909, p. 57).

The wires for the main cables were delivered in 24,384 m (80,000 ft) continuous length, wound on a reel. Four reels were placed at each anchorage, eight total on the bridge, allowing for eight traveling sheaves at a time (Hool and Kinne, 1943, p. 352).

Strands were supported at the anchorages and tower saddles by cast iron sheaves bolted temporarily to the saddles on each side of the groove, several inches above the tops of the saddles and 30.5 to 61 cm (1 to 2 ft) above their final position (Perry, 1909, p. 56). Movement of the traveling sheave was monitored. A system of electric bells and telephone notified controllers at the break wheels, greatly assisting all operations and adjustments (Perry, 1909).

A loop placing two wires was pulled by 91.5 cm (3 ft) diameter traveling sheaves, which made the round trip from anchorage at anchorage in 15 min. The traveling sheave on the opposite side for each cable also carried a loop, allowing placement of 16 wires at a time (Steinman, 1922, p. 339). Since the length of each cable is 983.3 m (3224 ft), the eight sheaves were laying wire at a rate of 64.4 km/h (40 mi/h). The 37,224 km (23,130 mi) of wire, 7% shy of the earth's circumference, were spun in less than 4 1/2 months—a record speed which inspired others to pursue more efficient spinning methods. For comparison, the amount of wire spun on the best day at the Brooklyn Bridge was 20 tons and 75 tons at the Williamsburg Bridge. The maximum amount of wire spun in one day on Manhattan Bridge was 130 tons (Steinman, 1922, p. 190).

Mayor McClellan was present at the start and end of the spinning, showing that Tammany Hall was capable of building public works in an efficient manner (Reier, 1977). He pulled the lever to lay the last wires on December 10, 1908. "As the wire was drawn over the Brooklyn tower, the spectators below cheered and passing river craft blew their whistles in salute. At the same time flags were unfurled on the towers of the bridges" (*New York Times*, December 11, 1908).

The Manhattan cables required the first hydraulic squeeze rings adaptable for different diameters: for the 7 strands in the first stage and for the entire 37 strands in the second stage. The method was replaced with flat band seizings on later bridges (Steinman, 1922, p. 340).

Holton D. Robinson was the Engineer in Charge of the Department of Bridges for the Manhattan Bridge in 1905 and worked for the contractor during the wire spinning. He designed and patented the cable-wrapping machine. This machine used an electric motor and was self-propelled for the first time. The 454 kg (1000 lb) wrapping machine used a 1.5 hp electric motor and pressed the wires against the preceding coil at 13 revolutions per minute with two spools at the same time advancing at a rate of 5.5 m/h (18 ft/h) (Steinman, 1922, p. 183; Hool and Kinne, 1943, p. 355). In 1921 Robinson and Steinman started a consulting firm.

The total length of the loaded cable between the pins of the anchor chain is 983.4 m (3226.35 ft) and for the unloaded anchor chains, it is 982.25 m (3222.61 ft). Thus, the extension due to the dead load of the trusses and floor is 1.14 m (3.74 ft) (Perry, 1909, p. 52). The lengths and dead-load forces were computed for parabolic curves.

Upon galvanization, the cable wires demonstrated outstanding ductility. They could bend cold around a rod 1.5 times their own diameter without signs of a fracture (Perry, 1909, p. 52). For protection from the weather and facilitation of handling and stringing, the wires were covered with grease during all operations (Perry, 1909, p. 66). The wire surfaces retain remnants of an oily coating 105 years later. In an early demonstration of sustainable economy, the 4.45 cm (1 3/4 in) footbridge cables were cut and used for the short suspenders (Perry, 1909).

1.1.4.4 Stiffening trusses

Manhattan was the first suspension bridge to use the lighter Warren truss. Erection proceeded at four separate points, simultaneously working from both directions of the each tower. The first pass was started in March 1909 and connected at midspan a little more than a month later (*New York Times*, December 5, 1909). In it, the lower chords of the truss and floor system were temporarily connected to the suspenders. The truss diagonals were installed on the second pass, followed by the upper decks and transverse bracing. For the trusses, 300 men were employed, erecting a record 300 tons per day (Steinman, 1922, p. 181). To achieve proper profile of the travel way and the cables, the chord members of the truss were closed only after the dead load was on the structure and adjustments were made to the suspender lengths.

The slender design of the bridge, apart from the deflection theory, is also due to incorporating nickel steel with a working stress of 2812 kg/cm² (40,000 psi) in the upper and lower truss chords. The working stress of the nickel rivets was 1406 kg/cm² (20,000 psi) (*Scientific American*, 1908). In all, 8100 tons of nickel steel was used for the trusses, and 44,000 tons of steel was used for the entire bridge (Steinman and Watson, 1941, p. 367).

Not half a dozen men lost their lives during construction (*New York Times*, December 5, 1909). The construction cost estimate in 1908 was US$26 million (Winpenny, 2004, p. 18; *Scientific American*, 1908) and the structure was completed for US$32 million.

1.1.4.5 Design and construction timeline

1897	Bills are introduced in the state legislature.
November 1898	The Board of Public Improvements authorizes preparation of the plans to construct an all-railroad bridge (New York City Department of Bridges, 1904).
November 1899	The Board on Public Improvements approves a wire cable suspension bridge.
January 29, 1900	Construction is approved by the municipal assembly, the mayor, and the War Department.
December 1900	Bids are opened for the foundation.
April 22, 1901	"Final bids for the construction of the Brooklyn Tower foundation for the bridge were opened, and the contract was awarded to John C. Rodgers, the lowest formal bidder, at a contract price of $US 471,757. This contract was executed on May 1, 1901, and the actual work was begun on August 20, 1901. The estimated cost of this structure, including approaches, is $US 15,833,600. The amount of money expended on this bridge to November 30, 1901, is $US 89,283.42" (Shea, 1901).
March 1902	Commissioner Lindenthal proposes changing the design to eyebar suspension bridge (Nichols, 1906, p. 23).
March 10, 1903	Eyebar substitute is successively approved and rejected by the Art Commission.
August 1904	The foundations of Manhattan and Brooklyn Towers are completed (New York City Department of Bridges, 1904, p. 22).
September 1904	Revised wire cables are approved by the Art Commission (New York City Department of Bridges, 1904).
August 1906	Phoenix begins receiving contracts for the superstructure between the towers (Winpenny, 2004, p. 11). Tower fabrication begins.

June 15, 1904	(New York City Department of Bridges, 1904, p. 22).
July 1908	Towers are completed (Hool and Kinne, 1943, p. 357).
June–July 1908	Temporary cables are strung and footbridge is constructed.
August–December 1908	All cables are laid (New York City Department of Bridges, 1912).
February–December 1909	Suspended-span steel is erected. Approaches are completed.
December 31, 1909	Official opening is held.
January 1913	The first coat of structural paint is applied.
1918	The bridge is fully opened to subways.

1.1.5 Traffic

Surface trolley started in 1912; light rapid transit, in 1915 (Steinman, 1955, p. 2). In 1918 the number of motorized vehicles in New York City surpassed that of horse-drawn vehicles. In July 1922 the Manhattan Bridge was closed to horse-drawn vehicles that had to cross via the Brooklyn Bridge (Winpenny, 2004, p. 40) and the Manhattan Bridge became much cleaner. In 1929 New York City Mayor Jimmy Walker bought the franchise ridges and rolling stock of the trolley line operating on the Bridge (south side), which was converted for bus use (Winpenny, 2004). Daily vehicular traffic increased from 65 000 to 110 000 cars (Winpenny, 2004). Pedestrians were banned from the East River bridges at night starting December 17, 1941 (Winpenny, 2004).

1.2 LIFE-CYCLE PERFORMANCE

The loads and responses are uniquely dynamic in suspension bridges. In that respect, they resemble mechanical devices more closely than they do "rigid" structures. With the Manhattan this is especially true, and as a result, the bridge has been a foremost subject of bridge management studies (Birdsall, 1971). The live loads on the bridge are exceptionally high in both frequency and amplitude and cause a similarly extreme structural response. To begin with, the bridge not only moves with seasonal and daily thermal changes but also can deflect over 0.3 m (1 ft) at center span every time one of the more than 900 daily trains crosses. This change in profile translates to frequent movement in the expansion joints at the towers and differential vertical movement most pronounced between the inner trusses. Figure 1.4 shows a 36 cm (14 in) sliding bearing under a subway track stringer, extended almost beyond its pedestal as a result of regular movements caused by service loads and temperature. Figure 1.5 illustrates a typical live load configuration producing the torsional response that has been questioned, investigated, and mitigated throughout the life of the bridge.

Figure 1.4 Overextended sliding bearing under the subway tracks.

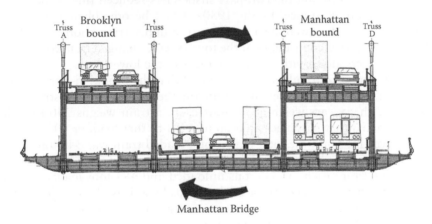

Manhattan Bridge

Figure 1.5 Typical cross section with asymmetric loading.

1.2.1 Torsion

"For reason unknown, Moisseiff placed the heaviest live loads, the subway trains and elevated-car traffic, outside rather than inside… resulting in excessive torque…" (Winpenny, 2004, p. ix). The effects of the differential deflection became apparent within 6 years of the bridge opening, as the bracings between the top chords of the inner trusses (B and C in Figure 1.5) cracked. Originally intended as support for traffic signs, the bracings were removed in 1924.

The three approved designs for the Manhattan Bridge may have contributed to a gap between the originally anticipated and the actual use of the bridge. Complicating this was the fact that the commissioner of New York City Department of Bridges was not in control of the transit facilities, as had been the case at the Brooklyn and Williamsburg Bridges. Under the new charter (*New York Times*, December 5, 1909), control had passed to the Public Service Commission, which succeeded the Rapid Transit Commission. Thus the bridge acquired a "split personality," responding to vehicular mode of transportation in a predominantly flexural mode of deformation and to rail traffic in a different mode, most significantly, a torsional one.

The original suspenders passed through the upper truss chords without centering devices. As each passage of one or two adjacent trains tilted the bridge to one side, the suspenders moved toward the upper chord webs, engaging in friction midspan. To improve the clearance and reduce chaffing against the upper chord, the original two-part suspenders were replaced with single-part suspenders near center span in 1937.

Measures mitigating the torsional movements were implemented between 1930 and 1940. In 1938, two-part suspenders near midspan replaced the lower portion of the four-part suspender ropes. Torsional displacements had caused damage to the original suspenders when they came in contact with the top chord and the two-part suspenders reduced this problem by providing more clearance. By the 1980s it had become obvious that the two-part suspenders were also engaging the upper truss chord. In 1938, the stiffening truss connections at the towers were changed from rockers to pinned hangers along with expansion joints in the lower deck.

Commissioner Zurmuhlen in 1952 noted that the Brooklyn–Manhattan Transit trains placed "a terrific strain on the cables and all structural parts." With six cars weighing 40 tons each, a train weighs 240 tons. It was estimated that passengers add up to 44% to that load, or 105 tons of human weight. Thus, the total weight of a loaded train was 345 tons. The trains crossing the bridge since the 1980s are composed of eight R68 cars, weighing 42 tons each, or 336 tons unloaded and 484 tons loaded. Thus, the new train loads exceed those of the 1950s by at least 40%. Added passenger capacity may have raised that ratio to as much as 60%.

In 1955 the city retained the firm Steinman, Boynton, Gronquist & London to conduct an extensive study of the bridge. As one option, Steinman's (1955) report recommended relocating the subway train to a new US$90 million tunnel. This was rejected by Mayor Wagner as financially impractical. The second option—to relocate three tracks to the center for US$30 million—was also never pursued due to operational difficulties. Many of the actions preceding the recent rehabilitation program have aimed at reducing the torsional rigidity, under the assumption that no stiffness is better than inadequate stiffness (Birdsall, 1971). Consistently with the original premise of the deflection theory, this logic produces excessively flexible structures. The 1955 study established that torsional stresses were responsible for cracks in the floor systems. The upper floor systems were replaced with a system supposedly less sensitive to twist during 1959 to 1962.

During a subsequent study in 1971, Steinman, Boynton, Gronquist & London tested a two-dimensional model at Columbia University in order to determine the spring constant of the cable truss system, acknowledging that the perfect dimensional model was the bridge itself. The study concluded that side-span supports, combined either with transverse stays at the main-span quarter points or with tower stays, and four panels of diagonals at the main-span centerline would most efficiently and economically reduce deflections and, hence, the torque. It was noted that the torsion had not compromised the performance of the primary members, namely, the cable, towers, and foundations (Birdsall, 1971). Several schemes to reduce the torsion were recommended, including tie cables and diagonals in the main and side spans and tower and transverse stays. None of these schemes were adopted.

The early 1980s' in-depth inspection found widespread and severe deterioration in the floor beams and stringers under the joints and at the fascias. Track bearings were worn out and laterals were broken. On at least one occasion in 1988, all nine stringers supporting the lower roadway were found broken at the same floor beam.

The central finding of the inspection was unexpected and serious: the upper roadways of the suspended spans, floor beams, and stringers, installed in 1962, had cracked extensively. Cracked members were clustered near anchorages, towers, and the center span. Figure 1.6 shows a typical crack in a stringer, propagating beyond a 2.5 mm hole, drilled in an attempt to arrest it. Figure 1.7a shows a crack through the entire bottom chord of truss D. A typical crack in the bearing of a stringer supporting the subway tracks is shown in Figure 1.7b. Advanced corrosion and poor bearing details contributed to an unquantifiable degree to the crack initiation and propagation.

New studies of potential stiffening schemes were conducted in two phases. As a foremost constraint, any structural modification had to increase the dead load on the cables and towers within acceptable margins of safety. In the first phase, a two-dimensional computer model analyzed 14 stiffening schemes, including those considered in 1955 and 1971. Figure 1.8a through c shows 11 of them.

Figure 1.6 Typical crack in roadway stringer, propagating beyond a hole.

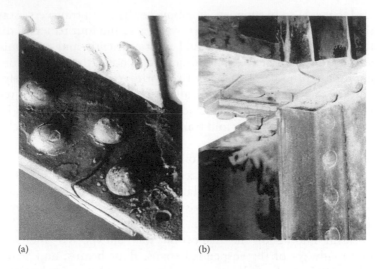

(a)   (b)

Figure 1.7 Cracks in (a) bottom chord, truss D and (b) stringer bearing.

In 1982 Weidlinger Associates first recommended creating what is called twin torque tubes to reduce the differential deflection between the trusses due to eccentric train loading. Three-dimensional models of schemes C1 and C2 (Figure 1.8) were studied analytically and validated through results of a load test. The high-friction forces in the stringer bearings of the upper roadways were identified as the key factor in causing widespread floor beam cracking. Failure of the stringers to slide over the floor beam could magnify the stress caused by the torque eightfold.

Scheme C2 was selected. It included stiffening the suspended torque tubes connected by strengthened floor beams at the lower level and providing eight rigid end frames around the transit envelopes adjacent to the towers for each suspended span, by installing diagonals between the outer and inner trusses of the upper level and diagonals between the outer trusses of the lower level. The partial upper laterals, removed in 1924, were replaced with stronger members. All sliding bearings under the stringers were replaced with elastomeric bearings, thus isolating the stringers from the torque tubes. Along with its structural benefits, the torque tube solution retained the appearance of the bridge. Each torque tube ends with a new rigid frame, as shown in Figure 1.9.

1.2.2 The stiffer performance

The torque tube and lower roadway stiffening system has performed under actual loadings only since 2008. The projected fatigue life of the stiffened structure has yet to be demonstrated. Meanwhile, its dynamic response is measureable.

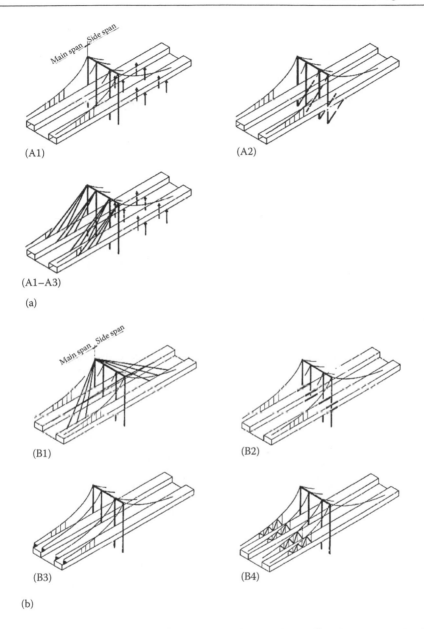

Figure 1.8 Stiffening schemes. (a) Category A: schemes that stiffen the category load; (A1) side span supports, (A2) underdeck cables in side spans, and (A1–A3) side span supports and stay cables in main span. (b) Category B: schemes that selectively stiffen the category load; (B1) crossed stay cables, (B2) interconnected hydraulic cylinders in truss chords, (B3) center span cable to truss connections, and (B4) diagonals between trusses and cables.

(Continued)

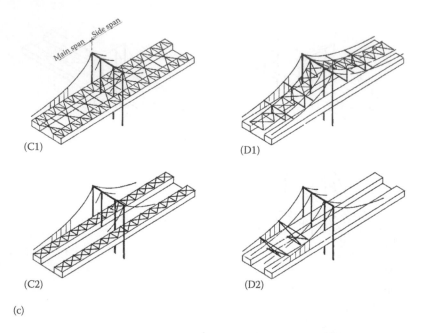

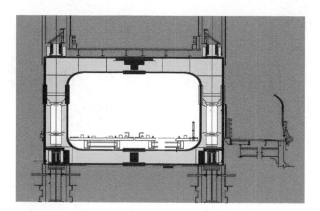

Figure 1.8 (Continued) Stiffening schemes. (c) Category C: schemes using torque tubes; (C1) full upper laterals and (C2) partial upper laterals and category D: schemes involving shear on the plane of the cable; (D1) torsional posts and (D2) transverse cables between suspenders.

Figure 1.9 Rigid frame constraining the end of each torque tube.

For the Steinman, Boynton, Gronquist & London (1971) study, Professor R. B. Testa of Columbia University measured up to 2.44 m (8 ft) in differential vertical movement between the bridge fasciae in the uneven loading case of two trains, on either the north or south tracks, as in Figure 1.5. Since the retrofit, measurements have been obtained by a laser tiltmeter

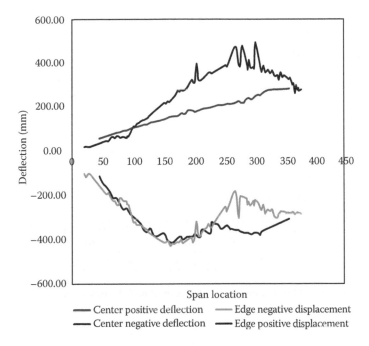

Figure 1.10 Laser tiltmeter measurements of deflection versus span location.

midspan (Figure 1.10), by an interferometric radar scanner by Ingegneria Dei Sistemi, Pisa, Italy (Figure 1.11), as a demonstration of the preceding system; and by a global positioning system (GPS) from the tops of the bridge towers (Figure 1.12), as part of the dynamic reanalysis of the bridge for seismic retrofitting (Mayer et al., 2010).

The three methods consistently indicate maximum differential movement of roughly 60 cm between the fasciae. Thus, the reduction in the original displacements by 50%, projected in 1981, has been exceeded.

1.2.3 Rehabilitation/reconstruction

By 1980, for structurally different reasons, the three great suspension bridges over the East River underwent major rehabilitation/reconstruction through multiple contracts. Between 1981 and 2020 the expenditures for the various capital projects on Brooklyn Bridge will reach US$936 million–US$956 million. The suspenders and stays, the decks, and some of the approach spans have been replaced. In 1988 the total replacement of the Williamsburg Bridge was considered. The cost of its rehabilitation/reconstruction (1983–2002) eventually amounted to US$1086.66 million. At Manhattan Bridge between 1999 and 2009, state and city inspectors

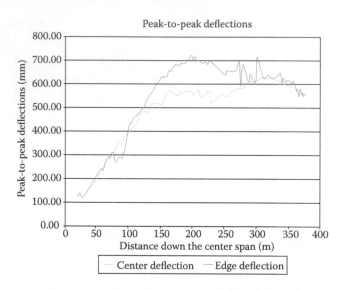

Figure 1.11 Results of interferometric radar scan of half the main span.

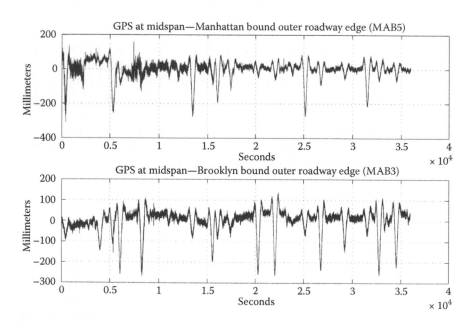

Figure 1.12 GPS receiver data showing outer roadway edge deflections.

issued 691 structural and safety flags. Of those, 206 were addressed by in-house forces and 485 were routed to various contracts. The following summary describes the scope and cost of the reconstruction/rehabilitation contracts from on Manhattan Bridge 1982 to 2012 (New York City Department of Transportation, 2012).

Rehabilitation items	Total estimated cost (million US$)
Repair of floor beams (1982)	0.70[a]
Replacement of inspection platforms, subway stringers on approach spans (1985)	6.30[a]
Installation of truss supports on suspended spans (1985)	0.50[a]
Partial rehabilitation of walkway (1989)	3.00[a]
Rehabilitation of truss hangers on east side of bridge (1989)	0.70[a]
Installation of antitorsional fix (side spans) and rehabilitation of upper roadway decks on approach spans on east side; replacement of drainage system on approach spans; installation of new lighting on entire upper roadways' east side, including purchase of fabricated material for west side of bridge (1989)	40.30[a]
Eyebar rehabilitation—Manhattan anchorage chamber C (1988)	12.20[a]
Replacement of maintenance platform in the suspended span (1982)	4.27[a]
Reconstruction of maintenance inspection platforms, including new rail and hanger systems and new electrical and mechanical systems; over 2000 interim repairs to structural steel support system of lower roadway for future functioning of roadway as a detour during later construction contracts (1992)	23.50[a]
Installment of antitorsional fix on west side (main and side spans) and west upper roadway decks; replacement of drainage systems on west suspended and approach spans; rehabilitation of walkway (installment of fencing and new lighting on west upper roadways and walkways); rehabilitation of cables in Brooklyn and Manhattan anchorage chambers; dehumidification of anchorages (1997)	141.82[a]
Installation of test panels (1982)	1.55[d]
Removal of existing suspender ropes and sockets in the suspended spans; replacement with new suspender ropes and sockets in the suspended spans and retensioning of suspender ropes bearing plates; retensioning of cable band bolts; removal of existing main cable wrapping; cleaning of main cables; new protective paste on main cables and new main cable wrapping; reinforcement of truss verticals and gusset plates; replacement of necklace lighting and multirotational bearings at trusses C and D; installation of access platforms at towers, rehabilitation of south upper roadway lighting (2010)	149.38[b]
Interim steel rehabilitation and painting—cable and saddle repairs lower roadway floor beams at panel point (PP) 37/38 on approaches and at anchorages; west side truss rockers and grillages on approaches; cable and suspender repairs; removal of parking deck; painting of entire west side, all four cables (2001)	127.98[a]

Rehabilitation items	Total estimated cost (million US$)
Stiffening of main span; reconstruction of North Subway framing; reconstruction of north upper roadway deck at suspended spans; rehabilitation of north approach span trusses; replacement of overlay on north upper roadway approach spans; rehabilitation of north elevated structures and subway tunnels; removal of railing on truss D in the north spans; painting of north side of bridge; new inspection platforms and debris protection in approach spans; construction of new north bikeway, replacement of approach span bearings and grillages; installation of intelligent vehicle highway system for north and south upper roadways as well as for lower roadway (in progress)	184.78[a]
Rehabilitation of lower roadway; rehabilitation of anchorage roofs under lower roadway; rehabilitation of substructures and retaining walls in Brooklyn and Manhattan approaches; installation of new signage on bridge and at plaza areas; installation of new lighting on lower roadway and plaza areas; cleaning and painting of lower roadway; installation of grating platform under towers at lower roadway; canopy lighting at towers (present)	143.80[a]
Seismic retrofit (2020)	40.00–60.00[c]
Total	880.78–900.78

[a] Complete.
[b] In construction.
[c] In design.
[d] Research and development (completed).

Major repair and construction of the torque tubes with full service closures were conducted at the south tracks from December 1990 to July 2001 and at the north tracks from July 2001 to February 2004. Figure 1.13 shows a new stiffening lateral member under the upper roadway. As during the original construction, the slip-critical bolted connections were tightened only after all laterals were installed.

Figure 1.13 New lateral bracing under the upper roadway.

Modular joints and finger joints had been alternatively used in the old roadways (Figure 1.14a and b). New modular joints are as shown in Figure 1.14c. A future transition to modular joints only is anticipated.

The new decks are steel grids (Figure 1.15) with concrete overfill. The entire 18,600 m² (200,000 ft²) of lower roadway deck, 2344 stringers, and 305 floor beams were replaced between October 2006 and August 2007 by introducing the following innovations:

- The work plan developed by the contractor (Koch Skanska) maximized access to the work zones by providing for delivery of equipment and materials from both the Manhattan and Brooklyn approaches. The construction began with two crews at the Manhattan anchorage with one crew working westerly toward the Manhattan abutment and one

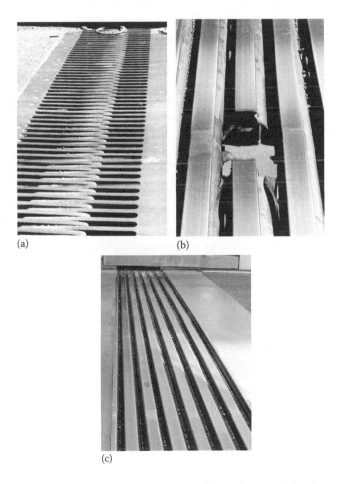

(a)   (b)

(c)

Figure 1.14 Roadway expansion joints: (a) finger, (b) modular, with broken spacer bar, and (c) new modular joint.

Figure 1.15 Lower roadway steel grid deck under construction.

crew working easterly toward the Brooklyn abutment. As a time-saving measure, the existing deck and stringers were removed in panels.

- The new stringers were preassembled in groups of two in the shop. The floor beams came to the site with the elastomeric pads preinstalled. This preassembly allowed for quick erection of the structural steel.
- To ensure that the contractor met the 12-month closure schedule, the contract specification required that 50% of the grid deck and structural steel be fabricated prior to closure. To minimize the risk, the contractor fabricated 100% of the main steel elements prior to the closure. The described measures not only lessened the impact on traffic but also improved the quality of the final product and reduced the duration of construction.
- Full closure of the lower roadway eliminated the need for construction joints in the grid deck and concrete placements were made from deck joint to deck joint—no cold joints were required.
- The grid deck panels (Figure 1.15) run the complete width of the roadway with no need for splicing of the main bars.
- The maximum incentive (US$65,000/day × 60 days) resulted in reopening the lower roadway 60 calendar days early.

All East River bridges were repainted with full containment, without traffic interruption. Seismic evaluation and retrofit are planned on all of them.

1.3 ANCHORAGES

Along with the towers, the anchorages render the cables load resistant and must precede their construction. The Manhattan Bridge ones are particularly monumental. With age, they became the center of uniquely innovative rehabilitation measures as well. By the early 1980s, the eyebars anchoring cable C in the Manhattan anchorage had lost so much cross-sectional area to corrosion that their elastic elongation was beginning to affect the tension of the strands. Weidlinger Associates designed the reanchoring of the cable, shown in Figure 1.16, and Karl Koch Construction was the contractor (Mayrbaurl and Good, 1988).

Two new transfer girders were anchored into the monolith by two drilled shafts, posttensioned and grouted. Nine strands were transferred to each girder (Figure 1.17). The sockets were filled with molten zinc. On a smaller scale, strands were subsequently reanchored in the Brooklyn anchorage, by using polyester raisin for the sockets. The anchorages are now actively dehumidified.

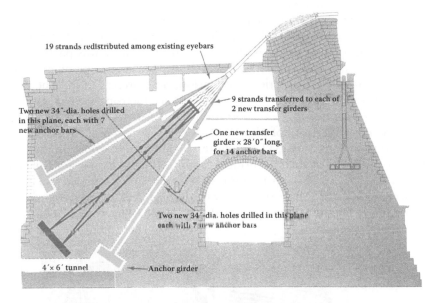

Figure 1.16 Manhattan anchorage, Manhattan Bridge.

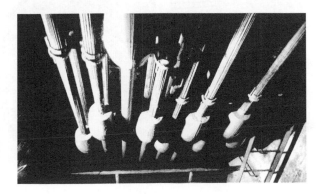

Figure 1.17 Nine transferred strands, cable C, Manhattan anchorage.

1.4 CABLES

Each of the four Manhattan Bridge cables comprises 9472 parallel high-strength galvanized wires (256 × 37 strands) with an overall diameter of 55.2 cm (21.2 in). Given a wire diameter of 5 mm (0.198 in), the void ratio of the cable is close to the typical, if not optimal, 20%. Figure 1.18a shows a portion of the cable, unwrapped and wedged for inspection. The cable wires are considered in good condition. Their red pigment is a remnant of the original lead paste. None was applied during the current rehabilitation. The original grease coating still covers the zinc surface of the wires. Figure 1.18b shows a test of the magnetic flux method developed by Tokyo Rope for estimating the amount of steel in the cable without unwrapping.

As the age of the parallel wire suspension cables surpasses a century, their life cycles enter unfamiliar territory. In 1988 the need to estimate the remaining strength of the Williamsburg Bridge cables was particularly urgent. That task intensified the research and development of methods for suspension cable inspection, evaluation, and preservation. The four suspension bridge owners in the New York City metropolitan area (New York City Department of Transportation [DOT], New York State Bridge Authority, Port Authority of New York & New Jersey, and Metropolitan Transportation Authority [MTA]) commissioned a joint report on the condition of the cables at their 10 bridges from Columbia University (Bieniek and Betti, 1998). The Transportation Research Board funded National Cooperative Highway Research Program Report 534 by Mayrbaurl and Camo (2004) on the evaluation of cable strength. The Federal Highway

(a)   (b)

Figure 1.18 (a) Unwrapped wedged cable and (b) testing of cable D by magnetic flux.

Administration (FHWA) followed with the report on field cable strength evaluation (Chavel and Leshko, 2009). The need to unwrap cables periodically, wedge them at numerous locations along their length and circumference, and examine the conditions of the wires was emphasized. Guidelines for statistical evaluation of the limited findings were recommended.

Concurrently, the noninvasive monitoring of cable wire condition by means of new technologies for data acquisition and transmittal has become a potentially cost-effective alternative to unwrapping and wedging. Through an act of Congress, FHWA funded an investigation of that possibility with particular emphasis on the East River bridges, as well as for general application. Many sensing technologies were tested on a 7 m (21 ft) 10,000-wire model at the Carleton Laboratory, Columbia University. The most promising ones were field-tested on Manhattan Bridge in 2012. The locations of the implanted sensors are shown in Figure 1.19. The cable model

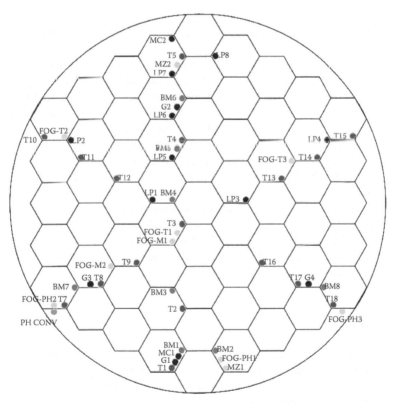

- ● T: Temperature/humidity sensors
- ● LP: Linear polarization resistance
- ● BM: Bi-metallic sensors
- ● MC: Couple multiple array (CMA) sensors (carbon steel)
- ○ MZ: CMA sensors (zinc)
- ● G: Granata sensors
- ○ FOG: fiber-optic sensors

Figure 1.19 Locations of various sensors in the cable model.

Figure 1.20 Cable model, Carleton Laboratory, Columbia University.

is shown in Figure 1.20. The results are reported in FHWA-HRT-14-024 (Betti et al., 2014).

The sensors installed with the assistance of New York City DOT at Manhattan Bridge monitored temperature, humidity, wind velocity, vibrations, and the rate of corrosion. A clear correlation was demonstrated between ambient temperature and internal humidity. That finding may explain why certain locations on a cable are more corrosion prone than others.

While noninvasive monitoring is explored in this manner, the Honshu–Shikoku Bridge Authority developed cable dehumidification by injection of dry air. The method precludes corrosion by maintaining humidity below 40%. The first applications were at the (then) new Kurushima and Akashi Kaikyo Bridges. The method is now contemplated for older structures, where linseed oil, lead, and zinc paste, as well as other active and passive corrosion inhibitors have been applied since the time of construction. The rate of air penetration under such conditions is investigated on the cable model at Columbia University (Figure 1.20).

1.5 SUSPENDERS

Cables were rewrapped and suspenders were replaced 30 years ago at the Brooklyn Bridge and 20 years ago at the Williamsburg Bridge. The suspenders at Manhattan Bridge had been replaced over 50 years ago. By the mid-1990 the suspenders were exhibiting wire breaks due to corrosion near the sockets at the level of the bottom truss chord (Figure 1.21) and due to friction at the level of the top chord.

Most of the existing suspenders on the Manhattan Bridge were installed under a US$2.2 million contract with Roebling & Sons in 1956 and was one of their last before closing their Bridge Division in 1964 (Zink and

Figure 1.21 Wire breaks in suspender at the bottom socket.

Hartman, 1992). That contract included an attempt to determine the dead-load stresses existing in the structure. Extensometers were used for the purpose with less than satisfactory results.

The latest contract to rehabilitate the main cables; install new wire wrapping, neoprene barrier, and hand ropes; and replace the 1256 individual suspender ropes at the 628 suspender panel points was bidden for US$149.5 million and executed between 2009 and 2013 by Koch Skanska.

As shown in Figure 1.22, the new suspenders represent a significant innovation. With the exception of the portions midspan and near the anchorages, the suspenders are anchored at the top chords of the trusses, rather than (as originally) at the bottom. The suspension reanchoring offered several life-cycle advantages. Due to the unique aspects of this new configuration, much

Figure 1.22 Old and new suspenders.

attention was given to the proper distribution of the loading on the suspenders. Frequency measurements were taken in the longitudinal and transverse directions at the equalizer bars of the new suspender assemblies with tri-axial accelerometers and initially calibrated with hydraulic jacks and pressure gauges. Loading of existing suspenders was measured by jacking prior to removal in order to calibrate current as-built loading of the main bridge.

The method considerably accelerated the project. The consultants (Weidlinger Associates and Parsons Transportation Group) also considered checking the new suspender loads using a "laser load" technique.

The geometry layout of the bridge and the requirement for only off-peak upper roadway lane outages dictated the replacements of the suspenders along each of the four cable lines before proceeding to the next.

1.6 LIFE-CYCLE MANAGEMENT

Manhattan and Williamsburg provide the only mutual rerouting alternatives for truck traffic between Brooklyn, Manhattan, and New Jersey. The capacity of the Brooklyn Bridge is limited to passenger and emergency vehicles. Hence, the concurrent rehabilitations had to be coordinated for minimum service interruptions. The very choice of rehabilitating, rather than replacing the Williamsburg Bridge, was influenced by the demand for uninterrupted traffic. Staging the rehabilitations of the East River bridges over the same 30 years elevated their costs to roughly US$1 billion per bridge. These expenditures, and the incalculable user costs, incurred during the inevitable traffic interruptions, underscored the need for cost-effective management of such unique assets over their entire life cycles. It was recognized that while any material structure and all of its modifications over time have finite useful life, the service provided by an essential bridge permanently changes the local geography and must have a perpetual life cycle. Similarly, perpetual is the need for structural maintenance and preservation, since by the conclusion of decade-long projects, new needs are already arising.

The developments at the East River bridges during the 1980s and 1990s contributed to the interest in life-cycle bridge management on all levels from the federal to the local. The following innovations in bridge management have resulted, among others, in the following:

- The FHWA recognized expenditures for major maintenance, such as repainting as eligible for federal funding.
- The New York City DOT commissioned a preventive maintenance report on the bridges in its purview (Bieniek et al., 1989, 1999). As a follow-up, detailed maintenance manuals were compiled for all East River bridges. Tasks such as oiling of wires in the anchorages, repairs of wrapping, painting, spot painting, maintenance of travelers, and cleaning of joints were scoped, scheduled, and budgeted.

- On the East River bridges and their approaches, deicing with salt by the city's Department of Sanitation was replaced by anti-icing with potassium acetate by New York City DOT. In a significant change of policy, the FHWA treats this costlier activity as eligible for federal funding.
- All designs for rehabilitation and replacement of structural elements must specify the expected useful life and supply maintenance instructions.
- Estimated life-cycle direct and user costs are gaining in importance over first direct costs as criteria in design proposal selection. For example, the reanchoring of the suspenders at Manhattan Bridge to the upper truss chords was motivated primarily by maintenance considerations.

1.6.1 Ownership and landmark status

Although the Manhattan Bridge is currently considered a city street, both New York City and State are currently looking for ways to collect revenue from the Manhattan Bridge, which is busiest trucking routes for the 7.7 million people living in Long Island (third most populous island in the Western Hemisphere). In the past, the city has suggested tolling through the PlaNYC congestion pricing. There was no vote on the proposal by the state legislators. New York State has suggested tolling through MTA's Blue Ribbon Panel. The MTA offered the city US$1.00 for the right to toll the bridge. A more equitable alternative proposes a fare-pricing plan that would toll all city and MTA bridges and tunnels to/from Long Island equally.

The Manhattan Bridge is an American Society of Civil Engineers National Engineering Landmark. The city's Landmarks Commission considers the portion of the bridge crossing four blocks of the recently designated Down under the Manhattan Bridge Overpass (DUMBO) Historic District part of their jurisdiction for any proposed work on the sub- or superstructure (New York City Landmarks Preservation Commission, 2009). Paradoxically, the Brooklyn anchorage may not be viewed by the Design Commission in the same way as the Manhattan anchorage was viewed by the Landmarks Commission. Consideration is being given to submit the entire bridge for City Landmark status to avoid the need to seek approval for two commissions for similar work on one contract.

REFERENCES

Betti, R., D. Khazem, M. Carlos, R. Gostautas, and P. Virmani, 2014. Corrosion Monitoring Research for City of New York Bridges, Report FHWA-HRT-14-023, Columbia University, New York.

Bieniek, M. et al., 1989. Preventive Maintenance Management System for the New York City Bridges, Report of a Consortium of Civil Engineering Departments of New York City Colleges and Universities, the Center of Infrastructure Studies, Columbia University, New York.

Bieniek, M. et al., 1999. Preventive Maintenance Management System for the New York City Bridges, Report of a Consortium of Civil Engineering Departments of New York City Colleges and Universities, the Center of Infrastructure Studies, Columbia University, New York.

Bieniek, M., and R. Betti, 1998. The Condition of Suspension Bridge Cables, Technical Report, Columbia University, New York.

Birdsall, B., 1971. Manhattan Bridge—Structure and Model: A Study in Motion and How to Arrest It, *Municipal Engineers Journal*, Paper 385, the City of New York.

Burr, W. H., 1913. *Suspension Bridges Arch Ribs and Cantilevers*, John Wiley & Sons, New York.

Chavel, B., and B. Leshko, 2009. Primer for the Inspection and Strength Evaluation of Suspension Bridge Cables, Federal Highway Administration, Washington, DC.

Griggis, Jr., F. E., 2008. The Manhattan Bridge: A Clash of the Titans, *Journal of Professional Issues in Engineering Education and Practice*, 134(3), 263–278.

Hool, G. A., and W. S. Kinne, 1943. *Movable and Long-Span Steel Bridges*, McGraw Hill, New York.

Johnson, A., March 23, 1910. Presentation to the Municipal Engineers, Paper 55, the City of New York.

Mayer, L., B. Yanev, L. D. Olson, and A. W. Smyth, 2010. Monitoring of the Manhattan Bridge for Vertical and Torsional Performance with GPS and Interferometric Radar Systems, Annual Conference, Transportation Research Board, Washington, DC.

Mayrbaurl, R. M., and S. Camo, 2004. Guidelines for Inspection and Strength Evaluation of Suspension Bridge Parallel-Wire Cables, National Cooperative Highway Research Program Report 534, Transportation Research Board, Washington, DC.

Mayrbaurl, R. M., and J. Good, 1988. Reanchoring a Main Cable on the Manhattan Bridge, The First Oleg Kerensky Conference, London.

Modjeski, R., 1909. Report on the Design and Construction of the Manhattan Bridge, For the Department of Transportation, New York City.

New York City Department of Bridges, 1901. Annual Report, Commissioner Gustav Lindenthal to the Mayor of the City of New York, Martin Brown Co., New York.

New York City Department of Bridges, 1904. Annual Report, Commissioner George McClellan to the Mayor of the City of New York, Martin Brown Co., New York.

New York City Department of Bridges, 1912. Annual Report, Commissioner Arthur O'Keeffe to the Chief Engineer Alexander Johnson, New York.

New York City Landmarks Preservation Commission, 2009. *Guide to New York City Landmarks*, Fourth Edition. Edited by Postal, M., John Wiley & Sons, Hoboken.

New York Times, December 11, 1908. Mayor Completes Last Bridge Strand, New York.

New York Times, December 5, 1909. One Hundred Cross the Manhattan Bridge, New York.

New York Times, January 1, 1910. Manhattan Bridge Opened to Traffic, New York.

Nichols, O. F., 1906. Manhattan Bridge, Its Terminals and Connections, Paper Number 23 presented March 27, 1906 to the Municipal Engineer of the City of New York.

Perry, A. I., January 14, 1909. The Manhattan Bridge Cables, Presentation to the Brooklyn Engineers' Club—Paper 86, New York.

Reier, S., 1977. *The Bridges of New York*, Dover Publications, New York.

Scientific American, February 1, 1908. Erection of the Manhattan Bridge Across the East River.

Shea, J., 1901. Report of the Commissioner of the New York City Department of Bridges to Honorable Mayor Robert A. Van Wyck.

Steinman, D. B., 1913. *Theory of Arches and Suspension Bridges*. Translation Authorized by Melan, J., Myron Clark Publishing, Chicago.

Steinman, D. B., 1922. *A Practical Treatise on Suspension Bridges: Their Design Construction and Erection*, John Wiley & Sons, New York.

Steinman, D. B., April 1955. An Investigation of the Structural Condition of the Manhattan Bridge, Report to the New York Department of Public Works.

Steinman, D. B., and S. R. Watson, 1941. *Bridges and Their Builders*, Putnam's Sons, New York.

Steinman, Boynton, Gronquist & London, April 1971. Study and Recommendations for Reduction of Torsional Deflections in the Suspended Spans, Manhattan Bridge, Report to the Transportation Administration, Department of Highways, City of New York.

Winpenny, T. R., 2004. *Manhattan Bridge: The Troubled Story of a New York Monument*, Canal History and Technology Press, Easton, in association with the Smithsonian Institution, Washington, DC.

Zink, C. W., and D. Hartman, 1992. *Spanning the Industrial Age: The John A. Roebling's Sons Company*, Trenton Roebling Community Development Corporation, Trenton.

Chapter 2

Akashi Kaikyo Bridge

Katsuya Ogihara

CONTENTS

2.1 BACKGROUND AND HISTORY

2.1.1 Japanese main island and strait crossings

The country of Japan consists of four main islands: from north to south, they are Hokkaido, Honshu, Shikoku, and Kyushu. Transportation by ships and airplanes between and among those islands has been often interrupted by bad weather, such as winds, waves, or fog. In addition, these means of transportation required longer travel time but provide less capacity. To eliminate such inconveniences and to promote balanced development of the country, construction of fixed links to connect those islands was carried out (Figure 2.1).

2.1.2 Honshu–Shikoku Bridge Project

In 1970, the Honshu–Shikoku Bridge Authority (HSBA) was founded based on the Honshu–Shikoku Bridge Authority Law as the sole project execution body. The Honshu–Shikoku Bridge Project includes three links between Honshu and Shikoku (Figure 2.2). In 2005, the Honshu–Shikoku Bridge Expressway Company Limited (HSBE) was established as the result of the privatization of its predecessor, the HSBA.

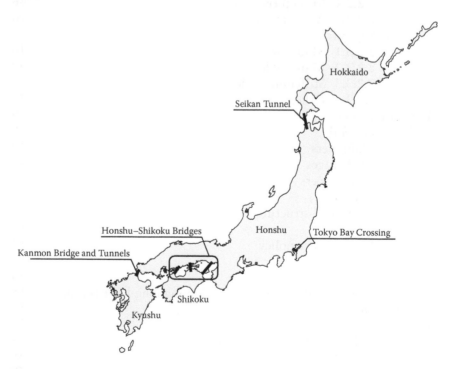

Figure 2.1 Strait crossings.

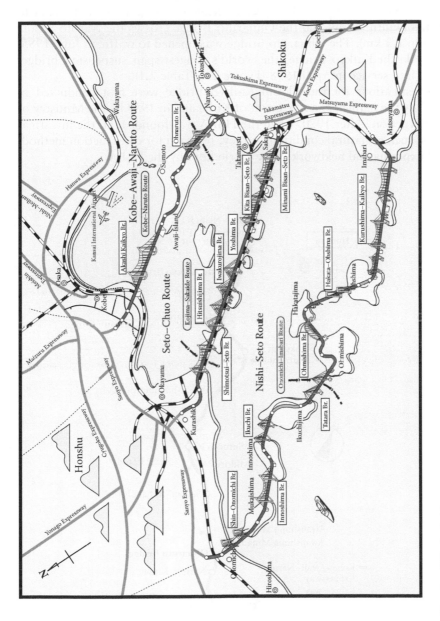

Figure 2.2 Honshu–Shikoku bridges.

2.1.3 Akashi Kaikyo Bridge

The eastern route of the Honshu–Shikoku bridges, the Kobe–Awaji–Naruto Expressway, is about 89 km long and runs from Kobe in Honshu to Naruto in Shikoku (Figure 2.3). The Akashi Kaikyo Bridge crosses the Akashi Straits (width: 4 km) and the Ohnaruto Bridge crosses the Naruto Straits (width: 1.3 km). The Ohnaruto Bridge was opened to traffic in June 1985. The Akashi Kaikyo Bridge, the world's longest-span suspension bridge, came into service in April 1998 (Figure 2.4; Table 2.1).

Investigations for the Akashi Kaikyo Bridge were first conducted in 1957 by the municipal government of Kobe and in 1959 by the Ministry of Construction. After 1970, when the HSBA was founded, all the investigations, such as natural conditions survey, research for construction method, and experimental fieldwork, were carried out by the HSBA.

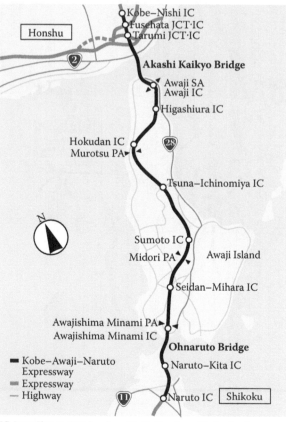

IC: Inter Change; JCT: Junction; PA: Parking Area; SA: Service Area

Figure 2.3 Kobe–Awaji–Naruto Expressway.

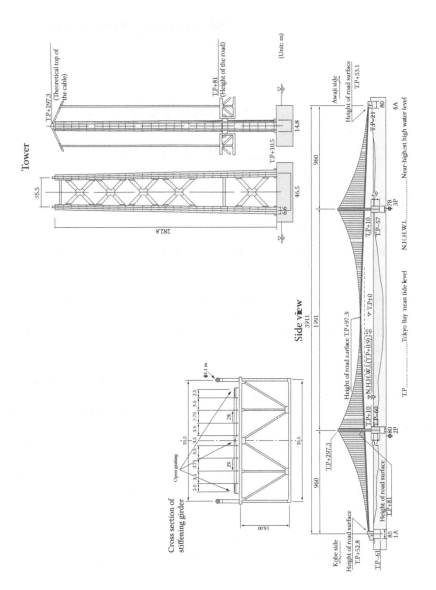

Figure 2.4 Dimensions of Akashi Kaikyo Bridge.

Table 2.1 Main dimensions of Akashi Kaikyo Bridge

Item	Dimension
Structure type	Three-span, two-hinged suspension bridge with steel truss-stiffened girder
Cable span	960 m + 1991 m + 960 m = 3911 m
Height of the road surface at the middle point of the center span	Approximately 97 m above sea level
Clearance beneath the girder	Approximately 65 m over the highest high tide
Tower height	Approximately 297 m above sea level (theoretical top of the cable)
Diameter and number of cables	$\phi 1.1$ m × 2
Height and width of stiffening truss	14.0 m × 35.5 m

2.2 DESIGN

2.2.1 Design condition

The basic design conditions of the Akashi Kaikyo Bridge are summarized as follows:

1. The width of the straits is about 4 km, and the maximum depth along the bridge route reaches about 110 m.
2. Natural conditions at the main pier site: Water depth is around 45 m, maximum tidal current is about 4.0 m/s, and maximum wave height is 9.4 m.
3. Wind condition: Basic wind speed for design (defined as 10 min average speed at 10 m above the water level with a return period of 150 years) is 46 m/s and the reference wind speed against flutter is 78 m/s.
4. Geological condition: The base rock beneath the straits is granite, on which the Kobe Formation (alternating layers of sandstone and mudstone in the Miocene), the Akashi Formation (semiconsolidated sand and gravel layers in the late Pliocene and the early Pleistocene), and the alluvium layer are deposited (Figure 2.5).
5. Design earthquake: This is an earthquake that occurs off the Pacific coast at a distance of about 150 km away with a magnitude of 8.5, or earthquakes that are expected within a radius of 300 km with a return period of 150 years.
6. Social conditions: These consist of a traffic route 1500 m in the width; heavy sea traffic is designated amid the straits, and lands on both shores are highly utilized. The bridge is for a six-lane highway with design speed for vehicles of 100 km/h.

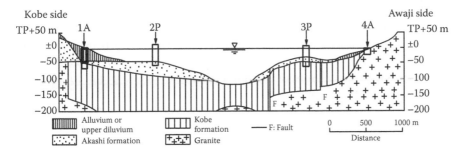

Figure 2.5 Geological conditions beneath Akashi Kaikyo Bridge.

2.2.2 Substructures

There are four substructures for the Akashi Kaikyo Bridge, two main piers and two anchorages, and they are called 1A, 2P, 3P, and 4A in sequence from the Kobe side. In this, *A* stands for anchorage and *P* stands for pier.

The minimum clear span length must be longer than 1500 m, which is the width of the statutory traffic route, and it is better for the main piers to keep off the traffic route with some margin, considering the usage of exclusive working areas during construction as means to ensure the safety of maritime traffic. Accordingly, several comparative designs were carried out for various main-span lengths of around 2000 m; the conclusion was that the span lengths of 1950 to 2050 m yielded the minimum construction cost. The positions of the two main piers were finally decided based on the topographical and geological conditions after securing the width of the traffic route and its side margins, which led to the main-span length of 1990 m (final length of 1991 m was due to influence of the Kobe earthquake).

1A is located at the reclaimed working area along the shoreline of Kobe side and has a foundation consisting of an underground circular retaining wall and roller compacted concrete (RCC) fill inside the wall. Its body, to which the two main cables are fixed, rests upon this foundation.

2P and 3P are circular underwater spread foundations constructed with laying-down caisson method.

4A, located at the Awaji side, is composed of a rectangular direct foundation and a body.

By considering the size of the foundations and relatively soft supporting ground, a new seismic design method was developed and used, in which concepts of "effective seismic motion" and "dynamic damping effect by interaction between foundation and ground" were introduced. Thanks to this new seismic design method and luck, the bridge withstood the Kobe earthquake in 1995 without receiving any structural damage.

2.2.2.1 Main piers

The position of 2P was determined at the edge of the sea plateau at a level of −40 to −50 m, from which point the seabed rapidly falls in depth. The Akashi Formation at −60 m was selected as the bearing layer for 2P.

The position of 3P was determined as a symmetric point to 2P with regard to the traffic route. This is −36 to −39 m deep, and the −57 m level of the Kobe Formation, the thickness of which is 40 to 65 m and which is partially exposed at the seabed, was selected as the bearing layer for 3P.

2.2.2.2 Anchorages

The side-span length was set at 960 m so that both anchorages might be located near the original shorelines. The body of two anchorages was designed to be a conventional gravity type. Because the anchorage body is a huge structure to which many people can make easy access, its shape was designed to lessen an oppressive impression, and precast panels, with their superior external appearance and pattern to avoid monotony of the concrete surface, were used. On the other hand, the two foundations were quite different in reflecting the difference of the geological condition. The 1A (Kobe side anchorage) foundation especially became a massive foundation with a diameter of 85 m and a depth of −61 m.

2.2.3 Superstructures

2.2.3.1 Main towers

The main tower is made of steel, and the shaft has a cruciform cross section, which is effective against wind-induced oscillation. The tower shaft was divided into 30 tiers and almost all tiers were composed of three blocks.

The cable reaction for the main tower is about 100,000 tons (1 GN). The dimensions of the tower shaft are 6.6 m × 10 m at the top and 6.6 m × 14.8 m at the bottom. The main steel grade used in the tower is SM570 (steel for welded structure with strength of 570 N/mm^2), and the total weight of one tower reaches about 24,700 tons.

Buckling stability was analyzed not only through linear buckling analysis but also through elastic–plastic finite deformation analysis, in which residual stress as well as initial deformation of the members was taken into consideration.

In order to reduce the amplitude of the vortex-induced oscillation caused by wind, not only during construction but also after completion, tuned mass dampers were installed inside the tower shafts.

2.2.3.2 Main cables

A cable is composed of 290 strands, with each strand containing 127 wires with a diameter of 5.23 mm. High-strength (1800 N/mm²) galvanized wire was developed and used for the main cables. The share of the dead load in the tensile force of the main cables of the bridge is much greater than that of other suspension bridges. This means the main cables of the bridge are safer against fluctuations in the live load compared to the main cables of other suspension bridges. Therefore, the safety factor for the allowable stress of the bridge was decreased. These measures helped avoid using double cables per side even when the span of the bridge was very long and the sag-to-span ratio was kept at 1/10. This sag-to-span ratio enabled restricting the height of the main towers.

One prefabricated parallel wire strand (PPWS) suspender rope consists of 85 galvanized steel wires with a diameter of 7 mm and a tensile strength of 1600 N/mm². The polyethylene tube of the suspender rope was colored with fluorine resin for aesthetic reasons.

2.2.3.3 Stiffening girder

Truss-stiffened girders were selected because the type offered advantages over streamlined box girders from the viewpoint of securing aerodynamic stability and ease of the erection to be done over the straits, which have rapid tidal currents and heavy sea traffic of 1400 ships per day excluding the smaller fishery boats. For precise evaluation of the aerodynamic stability of such a long-span bridge, a three-dimensional wind tunnel test using an aeroelastic bridge model was conducted in addition to the conventional two-dimensional wind tunnel test with rigid partial model.

Quenched and tempered high-strength steel, such as high–tensile strength steel with breaking strength of 780 N/mm² (HT780) and HT690, were used to reduce the dead load and to ensure sufficient strength.

The width and height of the main truss are 35.5 and 14.0 m, respectively, and these were determined in order to accommodate a six-lane deck as well as to have enough torsional rigidity for ensuring the aerodynamic stability. Two-hinged supporting condition was selected because the Akashi Kaikyo Bridge does not have a railway.

2.2.3.4 Others

The color green gray was chosen for the bridge. The bridge illumination was arranged, in which the color of cable illumination can be changed monthly or seasonally in accordance with a predetermined plan.

2.3 CONSTRUCTION

2.3.1 General

Record of the construction timeline for the Akashi Kaikyo Bridge is shown below:

April 1986	The official groundbreaking ceremony was held. (After that, final boring, excavation test, fabrication of caissons, and so on were conducted.)
May 1988	Actual work began at the site (reclamation for working area and excavation for main piers).
1989	The caissons were placed and underwater concreting began.
1990	The work on the anchorage foundations began.
1992	The main piers were completed and the tower erection was begun.
November 1993	The main towers were completed and the pilot rope was spanned.
1994	The anchorages were completed; the erection of the strands was begun and finished.
January 1995	The Kobe earthquake occurred. The bridge received no structural damage, but the position of the four foundations was slightly changed due to crustal movement by the earthquake.
June 1995	The erection of suspended structure was begun.
September 1996	The stiffening truss girder was closed.
April 5, 1998	The bridge was opened to traffic.

It is noteworthy that no fatal accident was recorded in approximately 10 years' work.

2.3.2 Main piers

Spread foundations, which were constructed with the laying-down caisson method whose shape was cylindrical with diameters of 80 (2P) and 78 m (3P), were chosen for main piers.

The HSBA had been familiar with laying-down caisson method since the construction of the Seto–Ohashi Bridges in the 1980s. However, the excavation method and underwater concreting method of the Akashi Kaikyo Bridge were improved from prior methods by using various technical developments.

In order to set the main piers on the bearing layer, the seabed had to be excavated by 15 to 20 m down to 60 (2P) and 57 m (3P) below the sea level. The required accuracy for excavation was decided as ±50 cm in order not to tilt the caisson that would be directly placed on the excavated seabed, and this requirement was met.

The steel caissons were cylindrical shapes with partially double-walled structures. The double-walled section, which was also provided with a bottom, generated enough buoyancy. These caissons were fabricated in dry docks and towed to the site. Positioning of the caissons was carried out with a controlling winch for each mooring rope. The allowable error for

Figure 2.6 Casting top slab concrete.

the placing operation was ±1.0 m. In the end, the caissons were successfully placed with an error of less than 20 cm.

As the Akashi Straits' scabed is covered with sand and gravel and its tidal current is rapid, there was the possibility that the caisson would overturn from the scouring caused by the accelerated flow and horseshoe vortexes, both of which are generated by presence of the caisson itself unless some countermeasures were taken. Accordingly, protection against scouring was provided with cobblestones (average weight of one piece was about 1 ton) and with filter units, which were fiber mesh containing smaller stones.

After the caissons were placed, newly developed underwater inseparable concrete filled the caisson. The concrete including the top slab was produced with a newly constructed concrete plant barge (Figure 2.6). A precooling system was equipped on the barge to control the temperature of fresh concrete to reduce the possibility of thermal cracking.

As for the cement, newly developed low heat–generating–type cement was widely used not only for the main piers but also for the anchorages.

The top slab concrete of the main piers, in which two tower anchor frames were placed, was ordinary reinforced concrete. Fiber-reinforced concrete panels were used for the side of the top slab in order to prevent carbonation of the concrete. The upper surface of the top slab concrete was also covered by polymer mortar against carbonation.

2.3.3 Anchorages

The construction procedure of the 1A foundation is summarized as follows:

1. Reclamation of the working area
2. Construction of a circular continuous underground wall by means of the slurry-trench method

3. Excavation of the inside soil and rock by using the underground wall as retaining wall while underground water is pumped up through deep wells

4. After excavation, filling the inside with RCC (a very hard, lean concrete that cannot be compacted with ordinary vibrators but with vibrating rollers) and then placing reinforced concrete of the top slab to complete the foundation

The geological condition of 4A is rather good because granite, which is the base rock of the straits, appears in shallow depths.

Each cable needs an anchor frame weighing about 2000 tons, which was fabricated in shipyards, from which the frame was transported by sea. Each frame was landed on the beachhead of the working area with a floating crane and then was moved to the designed position in backshore.

Precast panels with a total thickness of 25 cm were used as remaining form (not to be removed) for concreting. Highly workable concrete, which could flow as much as several meters by its own weight without segregation of coarse aggregate, was newly developed and used. As for cooling of the concrete to suppress the thermal cracking, both precooling and pipe-cooling methods were employed.

The total volume of this concrete was about 380,000 m^3 (140,000 m^3 for each body of 1A and 4A and another 100,000 m^3 for the foundation of 4A). Work was completed by April 1994 (Figure 2.7).

Figure 2.7 1A anchorage work.

2.3.4 Main towers

Almost all parts of the tower shaft (skin plates and stiffeners) are made of steel with a tensile strength of 570 N/mm², and the maximum thickness of the plate is 50 mm. The quality of the steel was strictly controlled according to the standards of the Honshu–Shikoku Bridge Material Code.

Half the load, which acts in joints of the tower shaft, is assumed to be directly transmitted by "metal touch" of the skin plates and plates and stiffeners, and the other half is considered to be transmitted by friction bolt connection. The accuracy of connection of each tier was strictly inspected at factories with thickness gauge of 0.04 mm, and the portion to which a thickness gauge could not be inserted (so-called metal touch ratio) constantly reached 90% to 100% of the entire surface area.

Each block was precisely fabricated in factories and transported to the site and then hoisted up for the erection with a self-climbing tower crane, which had a lifting capacity of 160 tons (1.6 MN) (Figure 2.8). High-tension bolt joints were used for the field connection. Final inclination at the top of towers after erection was measured as 1/7300 (specified allowable error: 1/5000) or less than 39 mm.

Figure 2.8 Erection of main tower.

The towers and the suspended structure were coated with newly developed fluorine resin paint that had high durability, and in this coating system, zinc-rich paint put directly on the steel surface plays an important role in anticorrosion performance.

2.3.5 Cables

There are basically two methods for erecting parallel wire cables for long-span suspension bridges: the air-spinning method and the PWS method. The PWS method was adopted. This method has been widely used in Japan because the method is less vulnerable against wind and needs less labor during the construction work.

A new method with a large helicopter was developed for spanning of the pilot rope. A light and strong polyaramid fiber rope (diameter: 10 mm; unit weight: 0.90 N/m; tensile strength: 46 kN) was used because the pulling capacity of the helicopter was not enough to carry a thick wire rope. An unreeler was hoisted down from the helicopter and operated by the crew so that the rope would not sag lower than 80 m above the sea surface. This 80 m was well above the determined clearance for the traffic route of 65 m.

Figure 2.9 shows the procedure for the cable installation.

The prefabricated strands, which were manufactured at a factory and wound around reels, were transported to the working area of 1A, which was the main base for all cable work.

The obtained final void ratio was 19.2% for ordinary section and 17.9% for the cable band portion, which were rather good results judged from the diameter of the cable.

2.3.6 Suspended structure

In the fabrication of the suspended structure of the Akashi Kaikyo Bridge, high–tensile strength steel, such as HT780, HT690, and SM570, was used mainly for the upper and lower chord members of the stiffening truss to better resist wind load without increasing the dead load.

One of the main reasons that the truss, not a streamlined box, was used for the stiffening girder was that erection could be done without depending on the sea surface under it; that is, erection in cantilever manner from the main towers was easy. Accordingly, the bulk of erection work was done in this method, while erection of some portions adjacent to the towers and anchorages was conducted with large-prefabricated-block method. Four blocks consisting of six panels, 84 m in length, were placed on both main-span side and side-span side of the towers (Figure 2.10). Two blocks consisting of eight panels, 112 m in length, were erected in front of 1A and 4A.

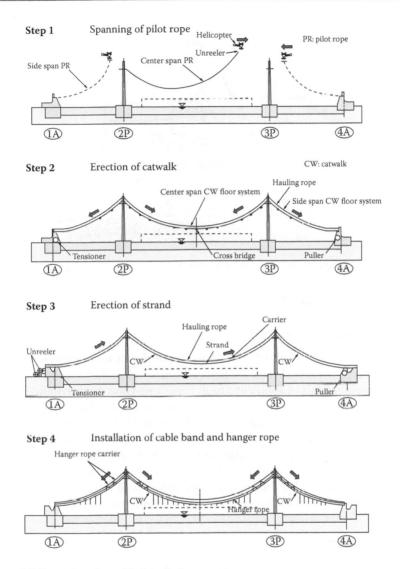

Figure 2.9 Procedure for cable installation.

The weight of each block at the work varied from 2700 tons (27 MN) to 3800 tons (38 MN). Facilities such as a crane and movable scaffolding for the erection were simultaneously placed on the block.

This erection method has been widely used in Japan because it could avoid disturbing the sea traffic lanes beneath the construction and had advantages in ensuring aerodynamic stability during erection.

Figure 2.10 Large block erection near main tower.

2.4 MAINTENANCE

2.4.1 Maintenance policy

Since the Honshu–Shikoku Bridges are part of the national expressway net-work, not only the Akashi Kaikyo Bridge but also all of the Honshu–Shikoku bridges must be kept in good condition for a long time. Maintenance works requiring extended closures of the expressway should be avoided to the utmost possible, because there are few alternative traffic routes. Therefore, preventive maintenance is adopted as a basic policy; appropriate inspections are conducted, deterioration predictions are conducted based on the inspec-tion results, and repair works are implemented before the performance of the structures degrades. Inspection, operation, and management systems have been established to carry out efficient maintenance (Figure 2.11).

2.4.2 Inspection

Inspections of the bridge structures are classified into regular inspection and unscheduled inspections. The regular inspection is composed of patrol inspections, basic inspections, and precise inspections. The patrol inspec-tion is implemented to find damage that affects the safety of the users and third persons. Frequency of patrol inspections is between once every 3 months to once a year. This inspection is implemented by visual checks from maintenance ways along the prescribed inspection routes. The basic inspection, a core part of the preventive maintenance, is implemented to grasp the changes in deterioration, to evaluate the degree of deterioration and to judge the need for countermeasures. Frequency of basic inspec-tions is between once a year and once every 5 years. These inspections

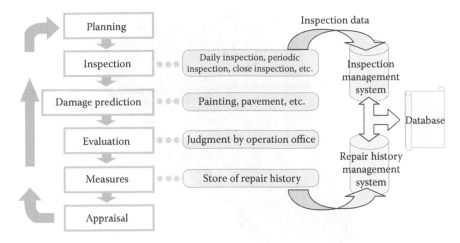

Figure 2.11 Flowchart of preventive maintenance.

are implemented in every detail by a visual check from maintenance vehicles. The precise inspection is implemented on the items considered of great importance on the safety of the entire bridge by using measuring instruments.

The unscheduled inspections consist of extraordinary-event inspections and extra inspections. The extraordinary-event inspection is implemented to judge needs for traffic closure and reopening in cases of extraordinary events, such as strong winds and earthquakes. The extra inspection is implemented on an as-needed basis to follow up on damage found by the scheduled inspection.

2.4.3 Maintenance vehicles

Since the structure is very high above sea level and many vehicles pass over the bridge and ships under it, many maintenance vehicles, such as maintenance vehicles for outside girder, for inside girder, and for cable, were installed to inspect the bridge safely and carefully (Figure 2.12).

2.4.4 Reducing life-cycle cost by adopting long-life painting

Painting the steel members of the Akashi Kaikyo Bridge, which is a long-life rust-proof type, consists of three layers. The base coat is inorganic zinc-rich paint, which includes rich zinc powder offering excellent anticorrosion performance by electrical and chemical sacrificial anode action. The undercoat, which protects the base coat, is epoxy resin paint, which has excellent adhesion and water and chemical resistance. In addition, fluorine

Figure 2.12 Maintenance vehicle for outside girder.

resin paint, whose performance is excellent against chemical action and weather action, is applied as the surface coat (Figure 2.13).

The HSBE has been trying to extend the recoating cycle by improving paint performance and works to reduce the life-cycle costs of painting. The maintenance concept for painting is keeping the base coat in sound condition. According to the concept, recoating work has to be completed until the top layer of the undercoat, which protects the base coat and the undercoat, wears off.

The site survey for the painting is carried out in order to check coat thickness, glossiness of the surface, coat adhesion, etc., every 5 years. Based on the trend of coat thickness surveyed, exposure time of the undercoat is predicted and the starting time of recoating is decided.

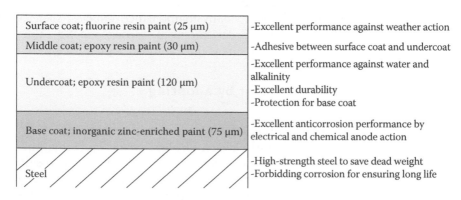

Figure 2.13 Long-life rust-proof–type paint.

Figure 2.14 Injection cover.

2.4.5 Dry air–injection system

The conventional anticorrosion method for the main cables of previous suspension bridges is a system composed of zinc galvanization of steel wires and prevention of seepage of rainwater into the cables with paste applied to the cables by using wrapping as well as painting.

However, surveys on several suspension bridges revealed that prevention of seepage of water was insufficient, and water and some rust was found on the wires. This shows that conventional anticorrosion methods are not adequate countermeasures for cables under the high humidity and wide-range temperature fluctuation conditions in Japan.

Since reerection of the main cable is almost impossible, a dry air–injection system was developed and installed to protect the main cable from corrosion by drying the inside of the main cable. The system, in operation in the Akashi Kaikyo Bridge since the time of the opening in 1998, was the first of its kind in the world (Figure 2.14). The target relative humidity is 40%.

2.5 OTHERS

2.5.1 Influence of the Kobe earthquake

In the early morning of January 17, 1995, an earthquake with a magnitude of 7.3 on the Richter scale occurred beneath the Akashi Straits. With an epicenter about 20 km west of downtown Kobe, it was very close to the Akashi Kaikyo Bridge, which was under construction at that time.

Consequently, the ground surface of this area deformed considerably. Surveys found the relative positions of anchorages and piers changed (Figure 2.15). The center span was elongated by about 80 cm and the side span near Awaji Island was elongated by about 30 cm. In addition, the

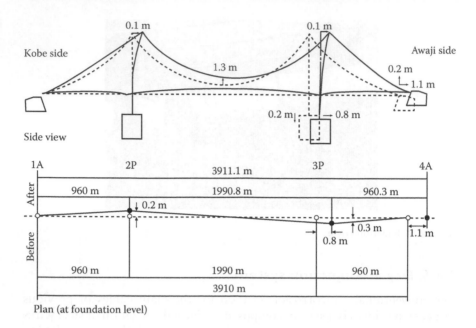

Figure 2.15 Relative deformation of Akashi Kaikyo Bridge.

centerline of the whole bridge was bent slightly. As a result, the center-span length was elongated from the design length of 1990 m to 1991 m. The length of the stiffening truss girder was adjusted at the sections, which were not yet fabricated.

2.5.2 Large boundary-layer wind tunnel

For precise evaluation of the aerodynamic stability of such a long-span bridge, it was considered necessary to conduct three-dimensional tests with an aero-elastic bridge model in addition to the conventional two-dimensional tests with a rigid partial model. The HSBA thus constructed a large boundary-layer wind tunnel (41 m wide, 30 m long, and 4 m high), which accommodated a 40 m long model (1/100 scale model) to verify the safety of the bridge. This was built in Tsukuba City near Tokyo (Figure 2.16). An outcome of this test was to establish the flutter analysis method, which could calculate the critical wind speed of flutter without depending on large sized wind tunnel tests.

2.5.3 Bridge-monitoring system

In the Akashi Kaikyo Bridge, various monitoring devices, such as a seismometer, anemometers, and an accelerometer, are installed and their data have been carefully recorded.

Figure 2.16 Large boundary-layer wind tunnel test.

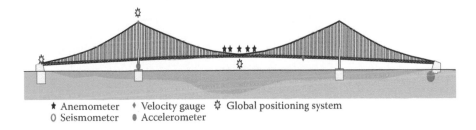

★ Anemometer   ♦ Velocity gauge   ☼ Global positioning system
0 Seismometer   ● Accelerometer

Figure 2.17 Monitoring system for Akashi Kaikyo Bridge.

Figure 2.17 shows the layout of the monitoring devices. The records are accumulated and analyzed to ensure structural safety by monitoring the behavior of the bridge. They can also be used to provide as precious information about the characteristics of nature, such as coherence and scale of the eddy of the gusty wind.

In addition to these systems, a GPS was introduced to monitor seasonal, daily, and hourly behaviors of the bridge, with factors that may be governed mainly by temperature and live loads.

Chapter 3

Tsing Ma Bridge

James D. Gibson

CONTENTS

3.1 INTRODUCTION

The Tsing Ma Bridge (TMB) links the islands of Tsing Yi and Ma Wan and is located within the Tsing Ma Control Area (TMCA), a 22 km transport network that provides both road and railway accesses from the urban areas of Hong Kong to the international airport at Chek Lap Kok on Lantau Island (Figure 3.1).

Figure 3.1 General view of TMB.

In the TMCA, there are two other landmark long-span cable-supported bridges (Figure 3.2), namely, the Kap Shui Mun Bridge (KSMB) and the Ting Kau Bridge (TKB). The KSMB is a double-deck cable-stayed bridge with a main span of 430 m and is linked to TMB, forming the Lantau Link. The TKB, a four-span, three-pylon cable-stayed bridge 1377 m in length,

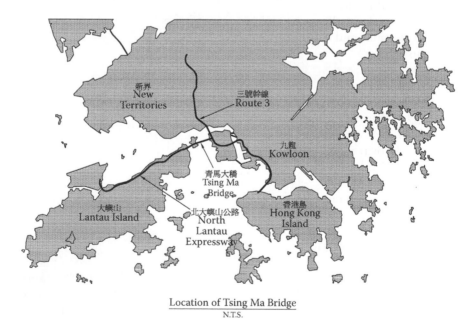

Location of Tsing Ma Bridge
N.T.S.

Figure 3.2 Hong Kong map.

forms part of the northbound route to mainland China and is linked to the Tsing Yi end of the TMB. The total construction cost of the TMCA facilities was US$2.7 billion and wholly funded by the government.

3.2 HISTORY

Feasibility studies were commenced in 1973 to investigate the possibility of expanding the usage of Lantau Island. This work concluded that construction of a fixed link would be financially viable and physically possible. A parallel investigation into the development of Hong Kong's aviation industry was also undertaken. This included the possible expansion of the existing airport or its relocation to a new site. Thirteen locations were examined and the selected best option was for a new airport to be constructed on or near Chek Lap Kok Island just off the northern Lantau shoreline. The Town Planning Board in 1976 further reviewed and consolidated the development schemes to provide for a new fixed crossing, a new urban development or town, and a new airport. In 1977, the government employed consultants to carry out more detailed feasibility studies into the three schemes and these were then further developed.

With regard to the fixed crossing, the study by Mott, Hay and Anderson (1982) focused on various environmental, geographical, and transportation factors, such as typhoon winds, currents within the shipping channels, ship impact, and types of traffic to be carried. It compared the various structural forms of bored tunnel, submerged tube, and bridge. After assessing such factors as capital, maintenance and operation costs, suitability for traffic, environmental impact, and shipping impact, a scheme incorporating long-span cable-supported bridges with high navigation clearances was adopted. At this stage the configuration of the main bridge was that the deck should be a hybrid truss structure, double decked with aerodynamic fairings and openings top and bottom. In terms of traffic-carrying layout, the bridge was to have two dual lanes on the upper deck and two single roadways and twin railway tracks on the lower deck. It was decided that the number of trains at any time on the bridge would be limited to two by means of appropriate signaling. It was also to have two large 900 mm diameter water mains, three 132 kV circuits, and nine telephone cables accommodated within the deck. The main span for this configuration was 1413 m, with an overall length of about 2050 m and shipping clearance of 62.1 m. The bridge was designed to be asymmetric with a suspended side span of 333 m on the Ma Wan end and straight stays overland on the Tsing Yi side, with the height of the towers to be 203 m (Figure 3.3). The detailed design was commenced in 1980 with all first drafts of contract documentation delivered in 1982 and the design checked in 1983. The estimated cost of constructing TMB was HK$1.5 billion on rates and prices effective in May 1982. Unfortunately the government chose to indefinitely postpone

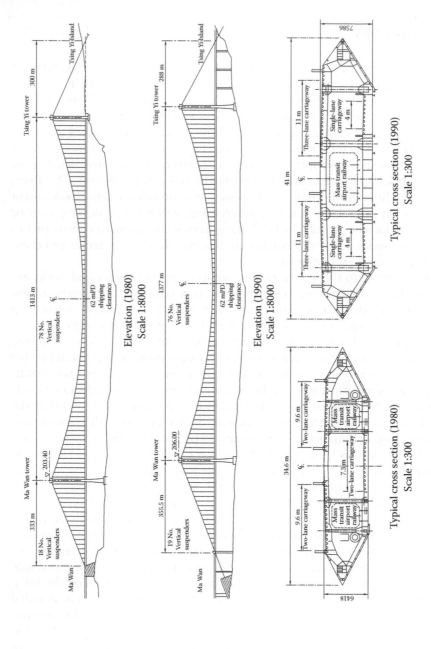

Figure 3.3 Comparison between 1980 and 1990 designs.

construction due to economic and other factors. If it had been built at this time, it would have been the longest suspension bridge in world.

In 1987, a new study was commenced called the Port and Airport Development Study. It was tasked with looking at the long-term provision of new port and airport facilities along with new urban developments and infrastructure. The outcome of this study was used to form the basis of a new strategic development plan and adopted in 1989. One key element of this plan was the Lantau Fixed Crossing, which resulted in a review of the previous scheme to update it to the new plan. The general route followed the original alignment; however, the crossing on landfall at Tsing Yi Island was moved further south, resulting in a slightly shorter main span of 1377 m, a 355 m suspended side span at Ma Wan, and the side span at Tsing Yi supported on piers with most of the other parameters remaining generally the same.

Detailed design was undertaken by the Mott MacDonald Group and completed by May 1992 when the first construction contracts were let.

3.3 DESIGN

The bridge was designed in accordance with British Standard BS 5400, as modified and supplemented by the Hong Kong Civil Engineering Manual, to achieve a design life of 120 years. Specific highway, railway, and wind loading plans were developed for the long-span bridge. In addition, special attention was paid to the corrosive atmosphere, marine environment, the requirements of the airport railway, and durability aspects of weld details and quality of concrete.

The design speed for the main highway was set to 100 km/h, while the railway was 140 km/h. Various combinations of traffic-loading criteria, such as loaded lengths, axle weights, dynamic effects caused by traffic jams, and different combinations of uses of upper and lower decks, were developed. For a loaded length of 2 km, the determined lane loading was found to be 14.85 kN/m. The railway loading was based on an eight-car four-axle train with static axle loading of 17 tonnes and a fatigue axle weight of 13 tonnes was used.

With respect to the design for the effects of wind, various wind records were examined to establish the most appropriate criteria. Three-second gust speeds of 80 and 85 m/s were selected for a 120- and 200-year return periods, respectively. For bridge operation under normal conditions, 3 s gust speeds of 44 and 50 m/s were used for highway and railway operations, respectively. Due to the occurrence of typhoons in Hong Kong, the bridge was considered to be without traffic during periods of maximum wind speeds.

A temperature range of ±23°C, a protection against ship impact by a 220,000–deadweight tonne vessel and a seismic peak ground acceleration

Figure 3.4 Bridge deck model.

of 0.05 g were adopted. Aerodynamic testing was carried out on various configurations of deck cross section. The longitudinal air vents were tested with combinations of differing widths and lengths along with varying leading edge profiles. The final deck section (Figure 3.4) consists of Vierendeel cross frames, supported on longitudinal diagonal braced trusses acting compositely with the orthotropic deck plates under the traffic running surfaces. The deck edge is clad in 1.5 mm thick ribbed stainless steel panels.

The incorporation of a railway onto the bridge created some further constraints to the design. A special rail movement joint was placed at the free end of the deck in order to accommodate the large longitudinal thermal and wind-induced movements. In order to maintain track alignment, lateral bearings were also installed to guide the bridge along its longitudinal axis. The loading from the rail cars largely dictated the design in that the fatigue life of the members became a significant factor and particular attention was paid to the adopted welding details.

Simple spread footings founded on competent rock were generally adopted for the foundation structures of the bridge, the only exception being Ma Wan tower, for which caissons were used due to the 12 m depth of water. Differential settlement between the tower legs was limited to 10 mm. Both anchorage structures are gravity structures. A safety factor of 2 against sliding was adopted for the foundations. The Ma Wan Island anchorage is partially buried, while the one at Tsing Yi Island is substantially below ground (Figures 3.5 and 3.6). The upper parts of these structures form the bridge abutments.

3.4 CONSTRUCTION

The main HK$7.2 billion contract for construction of TMB was awarded on May 25, 1992 to a joint venture consortium called Anglo Japanese Construction Joint Venture (AJC JV). It was composed of Trafalgar House Construction (Asia) Ltd. (from 1996 part of Kvaerner Group of Norway), Mitsui & Co. of Japan, and Costain Civil Engineering Ltd. of Great Britain.

Figure 3.5 Ma Wan anchorage excavation. (From the Highways Department, the Government of Hong Kong Special Administrative Region. With permission.)

3.4.1 Foundations for anchorages, towers, and piers

The geology of the area has been defined as being mainly a combination of intrusive and extrusive igneous rocks composed of tuffs with granites and rhyolites.

The Tsing Yi anchorage is located on the western flank of Tsing Yi Island. Firstly, the top of the slope had to be reduced down to a level of +58 m, with some 700,000 m³ of completely decomposed granite (CDG) and rock removed over 9 months. This material was then shipped across the channel and used to increase the land area on Ma Wan Island for construction purposes. The second stage was to excavate down to a level of –5 m mean principal datum, removing some 300,000 m³ of rock, and create the space (53 × 55 × 100 m³) within which to build the anchorage.

Figure 3.6 Tsing Yi anchorage excavation. (From the Highways Department, the Government of Hong Kong Special Administrative Region. With permission.)

During the massive excavation at Tsing Yi, the reclamation area on Ma Wan gradually increased. Within 8 months from commencement of the contract, the 80,000 m³ of excavation for the anchorage and Pier M1 was completed. Pier M2 had to wait a little longer for the reclamation to be further extended to encompass the work site. For the Ma Wan tower, it is situated offshore in an exposed location where it is not unknown for there to be white water conditions during the period of ebb tide. Firstly, the soft material was removed down to rock head, and then the rock was removed by underwater blasting down to approximately −14 m. In addition to the excavation, the surrounding area was partially backfilled in order to begin creation of the ship protection island; an opening was left to allow the tower leg base caissons to be floated in at a later date. Due to a large amount of overbreak caused by the very friable rock, precast pads were installed at the location of each caisson corner to ensure that the caisson grounded on line and level during their sinking. The caissons (Figure 3.7), 28 × 20 m² in plan and approximately 19 m high, were floated into place in April 1993 and handed over to the tower contractor in July 1993.

The foundation of the Tsing Yi tower consisted of a reinforced pad 27 × 19 m² in plan and 7 m deep, with rock located at −2 m. The other piers in the vicinity were constructed in a similar manner. For all reinforced concrete located below sea level, epoxy-coated reinforcement was used.

Figure 3.7 Ma Wan tower caissons. (From the Highways Department, the Government of Hong Kong Special Administrative Region. With permission.)

3.4.2 Towers

The two concrete towers each consist of twin legs joined together by four portal beams. The towers are approximately 210 m high and were constructed using the slip-form method. Three portal beams are located above deck level and one below. The shape of each leg follows a gentle curve from 18 m wide at its base in the longitudinal direction (in line with the deck) to 10 m wide at the level of the lower portal beam. From this point it tapers to a 9 m width until the last few meters, where it widens out to 13 m to form a seat for the cable saddles (Figure 3.8). The leg width in the transverse direction is a constant 6 m. Prior to construction, wind tunnel

Figure 3.8 Ma Wan anchorage splay saddle (tapered plates for splay). (From the Highways Department, the Government of Hong Kong Special Administrative Region. With permission.)

testing was carried out on a 1/200th model of the tower with cranes and other temporary works for various stages of construction. The resulting analysis indicated that there would be no need to have additional guy ropes for providing additional stability during the passage of a typhoon. For the portal beams (Figure 3.9), the analysis gave some critical slip-form levels, by which time the two lower portal beam trusses had to be erected and installed before slipforming could reach its full height. Each truss was raised by strand jacks and was accompanied by its reinforcement and formwork; the heaviest lift was 500 tonnes.

The Tsing Yi tower slipforming was completed in 110 days with 3 days being lost to bad weather. The Ma Wan tower was commenced 24 days before the completion of the Tsing Yi tower and was completed in 90 days with only 2 days lost. The main cable saddles (500 tonnes each and $11 \times 4.5 \times 6$ m^3) were then raised in three main sections by strand jacks, placed on a sliding bearing grillage plate with a 1.2 m offset toward the back span, and joined together. This permitted adjustments to be made to the saddles' position by jacks during the erection of the deck, in order to prevent too much bending of the leg overstressing the reinforced concrete.

3.4.3 Piers

The two Tsing Yi side-span piers closest to the tower (T2 and T3) were founded on rock with a pad 12×32 m^2 in plan and 3.5 m deep. The twin hollow legs (11×5 m^2) were constructed using the jump-form system to a height of 50 m. The third pier (T1) was located partly up the slope toward

Figure 3.9 Tsing Yi tower—raising of portal beam. (From the Highways Department, the Government of Hong Kong Special Administrative Region. With permission.)

the anchorage. It is a single 66 m wide reinforced cellular concrete structure due to the widening of the supported deck to include additional slip roads. The Ma Wan approach span piers are deliberately different from each other. Pier M1 (Figure 3.10) closely resembles T2 and T3 but is slightly shorter. However, pier M2 has more rounded, shorter faces corresponding to that of the main towers and this was done to emphasize that the suspended span begins. In addition, the side-span splay saddle is set on the pier top. Both piers were constructed by the jump-form system.

3.4.4 Suspension system

Cleveland Structural Engineering Ltd was awarded the contract by AJC JV for both the suspension system and the superstructure. Aerial spinning (Figure 3.11) was adopted instead of PPWS as a contractor's alternative

Figure 3.10 Ma Wan view of anchorage, pier M1, pier M2, and tower construction. (From the Highways Department, the Government of Hong Kong Special Administrative Region. With permission.)

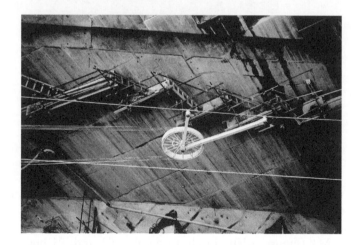

Figure 3.11 Spinning wheel on first pass at Ma Wan. (From the Highways Department, the Government of Hong Kong Special Administrative Region. With permission.)

proposal during tender stage. Each suspension cable is formed from 5.38 mm diameter wires with an ultimate limit state of 1570 N/mm², laid parallel to each other to have an approximate finished diameter of 1.1 m. In total there are 33,400 wires (91 strands) in the main span and 35,224 wires (97 strands) in each of the side spans, amounting to a total of 28,000 tonnes in both cables.

Figure 3.12 Cable compaction. (From the Highways Department, the Government of Hong Kong Special Administrative Region. With permission.)

A four-bight Roebling-type spinning system was adopted and used in order to accelerate wire placement, similar to the two-bight systems used previously on the Forth, Severn, and Bosphorus. After two strands had been spun, they were then lowered to their final position at night, when the temperature was more stable, by adjusting the strand shoes in the anchorage.

The design of the cable footbridge caused some initial difficulties in managing the loading under typhoon conditions. It was solved by ensuring that if a storm approached, then some of the storm strands (in a reverse catenary anchored low on the towers) would be tensioned to allow the footbridge to deflect and reduce the loads imposed on the eight main and two tram support strands. Spinning commenced in July 1994 with two 9 h shifts per 24 h and was completed in April 1995. The cable was then compacted with temporary straps installed to maintain the circular shape (Figure 3.12). One hundred and ninety cable bands were installed on the cable at 18 m centers, and two loops of 75 mm diameter steel wire rope suspenders were hung over each band, ready to support the steel deck. An additional 54 bands were also installed to secure the cable where no suspenders were present.

3.4.5 Deck manufacture and erection

Steel for the deck was supplied by both British and Japanese rolling mills and initially dispatched to fabrication shops in those countries and Dubai. There the steel was cut, welded, and assembled into small modules or elements and shipped to Hong Kong and China. The contractor set up an assembly yard at Shatian, Dongguan, some 80 km up the Pearl River and

Figure 3.13 Deck assembly in Shatian. (From the Highways Department, the Government of Hong Kong Special Administrative Region. With permission.)

over 60,000 m² in area, sufficient to assemble the required 96 suspended deck units, each 18 m long, 41 m wide, and 7.6 m high (Figure 3.13). The remaining lengths of deck would be tackled differently. The units were assembled from a number of elements, similar to a three-dimensional puzzle, into the final 18 m long sections. These were then laid out end to end and match fitted with their adjacent counterpart to ensure that the correct longitudinal camber was incorporated. The units were fitted out with as much of the secondary steelwork as possible and included the service walkways, emergency rail exit platforms, and stainless steel cladding.

In addition, the units were fully painted apart from the areas at the splice joints and the upper and lower deck traffic running surfaces grit blasted with the upper deck being zinc sprayed as well. To speed up work on site and to minimize the size of the strand-jacking crane, 88 out of 94 units were twinned together into 36 m long units. Two 36 m units were then moved onto a barge and transported down river on a 10 h journey. The lifting gantry (Figure 3.14) used strand jacks to lift each 1000-tonne unit. Erection (Figures 3.15 and 3.16) commenced in the middle of the main span and then proceeded on two fronts until 14 sections had been erected. Ma Wan side-span erection was then commenced. After 60% of units were erected, joining could take place. The first section was lifted on August 9, 1995, and the final unit on March 28, 1996, amounting to a total of 40,000 tonnes of steel rose in 8 months.

As parts of the approach spans were not only over land and at a height of 70 m but also above sloping ground, unusual methods were necessary for their erection.

Figure 3.14 Deck lifting gantry. (From the Highways Department, the Government of Hong Kong Special Administrative Region. With permission.)

For Tsing Yi the erection sequence was as follows:

- The span between T3 and T2 was assembled at ground level in two halves, then both halves lifted by strand jacks and slid across the pier tops and joined together. A derrick crane was placed on the first half.
- A 15 m length of deck was placed on the tower portal beam and another length of span between the tower and T3 was then lifted up and joined.
- For the remaining spans between T2 and the abutment, 15 m deck sections were raised by the derrick, successively joined with temporary props supporting the resulting cantilever.

At Ma Wan a modified approach was adopted as follows:

- Two 15 m lengths of deck were placed on the pier tops and the length of span between the piers was assembled and then lifted up in one piece and joined.

Figure 3.15 Tsing Yi side-span deck erection. (From the Highways Department, the Government of Hong Kong Special Administrative Region. With permission.)

Figure 3.16 Ma Wan side-span deck erection. (From the Highways Department, the Government of Hong Kong Special Administrative Region. With permission.)

- Two derrick cranes were placed on the deck.
- For the remaining spans between M1 and the abutment and M2 to the tower, 15 m deck sections were raised by the derricks and successively joined.

3.4.6 Railway

To control noise and vibration from trains running at up to 135 km/h, a special track-form arrangement was devised, composed of cross ties connected to longitudinal rail bearers made from steel boxes filled with concrete. This assembly was then supported on rubber bearings (9000# on Lantau Fixed Crossing [LFC]) and held in place by lateral restraints (6000#). The rails were connected in traditional fashion with chairs and clips. In addition, there were two guardrails to contain any derailment. Tests on various designs were undertaken in the United Kingdom on a test track prior to construction and later on the completed structure to ensure that the system was performing as intended. To facilitate disembarkation of passengers from a broken down train, platforms were installed along the full length of the crossing with exit gates to the lower deck roadway at regular intervals.

3.4.6.1 Bearings and movement joints

With a long-span bridge supporting road and rail traffic, the movements of the structure caused by loading, temperature, and effect of wind can create some very large displacements to the bridge deck. These are presented in Table 3.1. All of these movements have to be allowed for and the structure restrained and guided as it passes over and through various supporting structures, such as piers, towers, and abutments.

The deck is fixed at the Ma Wan abutment by four 310 mm steel pins that permit rotation. For the remaining part of the crossing, the deck is supported on a majority of sliding bearings on the pier tops and tower portal beams and has a series of road and rail movement joints at Tsing Yi. At Pier M2 and Ma Wan tower, the deck is supported and restrained on rocker or link bearings to resist uplift from certain traffic loading conditions. In addition, at each of the Ma Wan abutment, pier M2, both

Table 3.1 Maximum movements

Maximum movement	Displacement
Longitudinal due to thermal	±835 mm
Lateral due to wind	4.4 m
Vertical due to vehicle loading	4.7 m
Vertical due to temperature	1.3 m

main towers, pier T1, and the Tsing Yi abutment, there are bearings to resist the lateral loads.

The road movement joints are of the form of 24 lamella beams supported on joists and connected with an arrangement of control springs, with the gaps between lamellas being sealed with elastomeric seals that contain a textile reinforcement insert.

A very special rail movement joint had to be designed to accommodate deck movements of ±1 m and train speeds of 135 km/h. The rails from the bridge terminate in sets of tapered castings that form the expansion sleeve, which then guide the ends of the rails projecting from the abutment. These rails are supported on three sets of connected sledges that maintain support during bridge movements. Underneath there is a spline girder that takes out small deflections and rotations of the approach span deck. This is connected to a stiff support girder that carries the expansion sleeve and the sledges, supported on rails secured to the abutment follow on.

3.5 OPERATIONS AND MAINTENANCE

An operator is employed by the government of the Hong Kong Special Administrative Region (SAR) under a Management Operation and Maintenance Agreement (MOM) to manage both the assets and operate the facilities, normally for a 6-year term. To date, one company has been successful in securing all MOMs.

The operator is paid a fixed fee for the operation and maintenance through a deduction of the tolls collected from vehicles using the road network. The ownership of the control area still rests with the government. The agreement was formulated in such a way as to reduce the risk to the operator by clearly stipulating the obligations and duties to be performed and to provide a mechanism to manage the costs of unforeseen or nonscheduled works. This method has the advantage of keeping the tender prices low rather than being inflated due to unknown risks.

In addition, a government monitoring team (GMT) was set up to monitor the performance of the operator and ensure that the MOM provisions are fulfilled. The main responsibilities of GMT are to

- Approve the operator's proposed maintenance procedures and maintenance program;
- Monitor and carry out random checks on the operator's works to ensure compliance with the contract requirements; and
- Examine and monitor implementation of the operator's proposal for nonscheduled works (i.e., additional works).

3.6 INSPECTION

Scheduled inspections of TMB take one of three forms: a number of general inspections (GIs) in accordance with the governments schedule, which can be a mix of visual and close visuals over the contract period; a principal inspection (PI), close visual of every element once per 6 years; and safety inspections (SIs), weekly or monthly visual inspections on foot or by vehicle. The majority of inspections are performed by the operator's own staff, who have long-span bridge experience and a variety of inspection skills, some with nondestructive testing for steel and others for working at height on the concrete structures.

The intention and aim is to locate any defects prior to serious damage or deterioration occurring. On some assets, a representative sample is inspected; if a defined level or number of defects is identified, then an additional localized portion is subjected to further scrutiny (see Table 3.2 for examples). The scope and frequency can be reviewed annually and modified by the government on the basis of the previous inspection reports and the identified defects.

In addition to the normal routine inspections, from time to time, the operator is obliged to undertake SIs when some anomaly or significant event has occurred. The scope of the inspection is dependent on the event involved and the elements that may have been affected. Two examples would be the following:

Table 3.2 Examples of GI schedule

Element	Frequency	Nature	%
Vertical bearings	Every 6 months	CVI	100
Lateral bearings	Yearly	CVI	100
Movement joints—underside	Every 6 months	CVI	100
Main cable—between bands	Yearly	CVI	50
Main cable—wires within anchorages	Yearly	CVI/VI	50/50
Suspenders—main lengths	Yearly	CVI/VI	5/95
Towers—external surfaces	Every 2 years	CVI	50
Piers—internal surfaces	Every 2 years	VI	100
Piers—external surfaces	Every 2 years	VI	100
Orthotropic deck—upper	Yearly	CVI/VI	10/40
Railway beams	Yearly	CVI/VI	10/40
Longitudinal trusses—outer	Yearly	CVI/VI	10/40
Seabed sonar survey for scouring	Every 2 years	Sonar	100

Note: CVI, close visual inspection; VI, visual inspection.

- After a storm where the wind speed has exceeded 80 km/h, the scope would include wind barriers, safety barriers, suspension system, cladding, signage, marine elements, movement joints, and bearings.
- Defects or anomalies (including weld crack indications) arising from accidental damage or found during a routine inspection and a more detailed investigation or inspection is considered prudent or necessary.

Additional testing requirements are incorporated in the scope of the PI using nondestructive testing techniques. The tests most commonly carried out are as follows:

- Concrete: Schmidt (rebound) hammer, carbonation, half-cell potentiometer, and chloride and sulfate contents
- Steel: paint thickness and paint adhesion
- Cables: cable force measurement

3.7 EVALUATION

Numerous sensors were installed throughout the bridge at the time of construction to enable the monitoring of the health and performance of the structure. This system is known as the Wind and Structural Health Monitoring System. In general, there are temperature sensors located in the main cables, decks, and towers; strain gauges in both longitudinal and cross members (including orthotropic deck); accelerometers (both portable and static); anemometers; displacement sensors; a weigh-in-motion system; and a GPS for major element displacement. Data are acquired at one of a series of out station computers and then transmitted to a central computer for processing and further trend analysis. Once per MOM, a load test is undertaken as a check or validation of the computer model used for analysis of the bridge. All information related to this system is managed and controlled by the Highways Department of the Hong Kong SAR.

In addition to the above, there is a further supervisory control and data acquisition system that provides information such as the deck level/tower top wind speeds. A configuration inspection is undertaken once every 6 years, the purpose of which is to provide an independent record of the position (time and level) of certain fixed known points on the bridge and then compare them with previous results. Once obtained, the results are reviewed to identify trends in movement of key elements that may indicate some anomalies in tower verticality, alignment profile, and displacement of the main bridge components.

A partial acoustic monitoring system has just been commissioned to monitor the Ma Wan back-span cables for wire breaks. None have been detected so far.

3.8 MAINTENANCE

Under the MOM, there are two types of maintenance, scheduled and nonscheduled.

Scheduled maintenance is part of the operator's obligations under the MOM and includes the following:

1. Maintaining the cleanliness of the sliding surfaces of the bearings
2. Repair of accident damage to wire rope safety barriers
3. Checking the bolt tension and undertaking bolt tightening of the cable band bolts of TMB, one round every 6 years
4. Removal of minor objects trapped in the road movement joints
5. Removal of water that has accumulated or is trapped in parts of the bridge structure
6. General cleansing of the carriageways, stainless steel fairings, mesh, and fencing on the lower deck and removal of litter and debris

The operator is also obliged under the MOM to provide a paint gang of eight painters. The scope of this service is limited to repairs of defects that can be reasonably handled by the said gangs up to a quoted area within an agreed program. This service is also deemed to include the cost of paint, labor, plant, access, nondestructive testing, etc.

Nonscheduled maintenance includes works outside the MOM and might cover the following:

- Major repainting of a section of steel deck or the main cable
- Repair of concrete cracks and the treatment of corroded reinforcement
- Resurfacing or repairs of the carriageway
- Replacement of bearings or parts
- Replacement of movement-joint control springs
- Additional access platforms for inspection
- Application of a silane paste to the towers of TMB

The last item above requires further explanation. Due to the rather humid nature of Hong Kong's climate at certain times of the year and the levels of pollution generally experienced in the Pearl River Delta, the mix of pollution and moisture provides an ideal breeding environment for mould to grow on concrete surfaces. This growth appears as very unsightly dark staining. In order to remove the mould and restrict its return, the concrete is first cleaned by water jetting and once dry, a silane paste is applied to severely limit the retention of water on the structure, thus preventing reoccurrence of the mould. The usual effectiveness of this treatment is around 9 years.

3.9 BEST PRACTICES

3.9.1 Paint

The paint systems for the steel deck and main cable are presented in Table 3.3. For the main deck steelwork, the current condition of the system is good, with only very minor small areas of localized breakdown that require patch painting with a maintenance paint system. The main cable is in a different condition. The paint system is now over 16 years old and is showing signs of deterioration. A trial was undertaken on two cable panels to ascertain any differences in cost, appearance, and durability between a two-pack epoxy surface-tolerant–based system and a water-based styrene acrylic elastomeric coating system with a polyurethane finish coat. No major differences were found and a contract was let to paint five panels on both cables at midspan. The elastomeric system was chosen by the

Table 3.3 Paint systems

Paint type	Dry-film thickness (μm)
General surfaces steel deck	
1. Two coats of zinc phosphate HB primer	2 × 50
2. Two coats of micaceous iron oxide HB epoxy undercoat	2 × 150
Total DFT	400
3. If exposed to direct sunlight, top two coats reduced to 125 μm and additional two coats of polyurethane two-pack finish applied; to achieve opacity MDFT of 200 may be required	40–50
Main cable	
1. Red lead paste, followed by galvanized cable wrapping wire and two-pack etch primer	
2. Three coats of zinc phosphate epoxy ester undercoat	3 × 55
3. Micaceous iron oxide phenolic undercoat and finish	3 × 45
Total DFT	300
Suspender ropes	
1. Aluminum paste and metallic lead paste	Liberal coating by gloved hand

Note: DFT, dry-film thickness; HB, high-build; MDFT, minimum dry-film thickness.

contractor and this will be adopted for use on the other areas of main cable, which should receive a full repaint in the coming years.

3.9.2 Movement joints

Over the last couple of years, the modular lamella–type movement joints have required some maintenance. The original seals had suffered from punctures, resulting in water and debris penetrating the internal area of the joints. It was decided to replace a few of the badly damaged seals with an unreinforced elastomeric insert. As a consequence, additional restraining straps had to be fitted to the underside of the joints to maintain the maximum gap-opening limit of 65 mm. It was also decided at the same time to take the opportunity to increase the size of the guide frames and replace control springs and bearings, as some of the debris was beginning to have an impact on the performance of the joints.

3.9.3 Bearings

One main problem has occurred with bridge vertical bearings, related to the premature wear and extrusion of the polytetrafluoroethylene (PTFE) disks that allow the bearings to slide on the sliding plate set on top of the piers (Figure 3.17). The actual cause has not been investigated; however, a similar phenomenon was experienced on a few steel bridges in Europe and on some North Sea offshore oil platforms in the late 1990s and early 2000s. It is believed to be a case of a buildup of very large accumulated movements by a number of rapid and small but frequent movements, in all likelihood caused by the passage of the airport rail trains. The effect of these movements causes a rapid dispersion of grease and a consequential

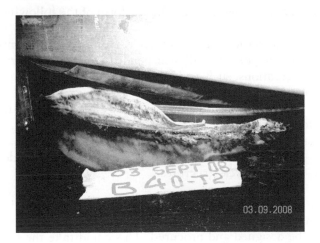

Figure 3.17 PTFE extrusion.

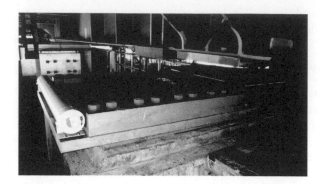

Figure 3.18 New bearings.

increase in friction. One secondary side effect of this is that the spherical bearings tend to rotate or spin, causing a misalignment of grease dimples, thus accelerating the wear effect.

As a result of the problems in Europe, the European bearing manufacturers actively investigated PTFE-replacement materials, which are more durable and have a higher resistance to wear without loss or undue increase in the materials' coefficient of friction.

In February 2007, a trial took place with bearings (Figure 3.18) incorporating a new sliding material, Maurer Sliding Material. Negligible wear and no extrusion have been recorded during the subsequent inspections to these bearings. Since then, all vertical bearings (22 in total) at the Tsing Yi end of the bridge where the maximum movements occur have been replaced.

3.9.4 Bridge access

In general, all new long-span bridges usually have various access facilities incorporated into their design and take the form of the following:

- Permanent underdeck gantries/travelers for bridge soffit inspection
- Permanent/temporary cradles/suspended working platforms in various configurations, such as curved and straight units
- Gondolas for main cables and suspenders
- Access ladders and platforms distributed throughout the main structures for general access
- Internal lifts in tower legs
- Other—mobile hydraulic hoist platforms used to provide access to above- or below-deck structures

There are always going to be shortcomings with some of the provided access. Enhancement, modification, and/or addition of new means of access will need to be designed and procured. This could take the form of temporary steel scaffolds for temporary use, purpose-designed platforms, or

Figure 3.19 Bridge-inspection vehicle.

the procurement of bridge-inspection equipment, such as cradles, hydraulic platforms, or bridge-inspection vehicles for permanent access (Figure 3.19). The basic design philosophy should be to ensure that the bridge inspector can get to most places by the simplest and safest methods. One example could be a high-level walkway installed at high-level areas to allow easy access to view the initial structural condition or to act as a starting point for a scaffold platform. A second could be to provide permanent platforms to those areas where access is required on a regular basis due to the criticality of the element, such as main cable wires in the anchorage, or for maintenance needs, such as the underside of wide expansion joints. It is always easier to incorporate access into the structure during the design stage.

On TMB, the underdeck gantries extend only part way up the underside of the external fairing. So for paint repairs to the splitter rail at the apex of the fairing, a special-purpose hydraulic platform is used. This inspection vehicle is provided by the Hong Kong SAR government for sole use by the operator. Despite the machine being capable of setting itself up within a single closed lane, progress is still rather slow.

Lastly, with a combined road and rail bridge, the safety and operational needs of the railway company does have an impact on the ability of the bridge operator to undertake inspections and maintenance such as painting. The usual problems are

- Time, i.e., short railway possessions that restrict the amount of time in which to complete tasks to approximately 4 h;
- Access to the areas above the overhead catenary power cables; and
- Work on areas of the structure immediately adjacent to the track.

3.10 FUTURE PLANS

As mentioned above, the life of the paint system in certain areas is showing signs of degradation and this will have to be attended to. The main cable will take priority and is to be recoated in phases over the next few years.

The existing lateral bearings are made up of pairs of spherical bearings (12 in total) bolted to the sides of the deck and mounted on precompressed rubber blocks behind a sliding plate incorporating PTFE. The combination of the different elements, with differing amounts of freedom of movement, causes the bearing assembly to assume a distorted twisted position. This is now being exacerbated by the gradual deterioration in the elasticity of the rubber blocks. To undertake any refurbishment of these bearings, the entire assembly will have to be removed, so planning is now taking place for this event. The mastic asphalt surfacing, which is only 38 mm thick, has performed quite well so far with only a few very small localized failures. However, it is now becoming apparent that due to driving habits, the volumes of certain types of vehicle, and the layout of traffic routes at the ends of the bridge, particular traffic lanes are suffering from different degrees of rutting. Again, plans are being prepared for some major resurfacing of parts of the bridge deck. This will involve closure of at least two lanes for the works and the running of traffic on the lower deck.

Early study has commenced for installation of a cable dehumidification system at some point in the future. There are currently no plans at present to undertake an intrusive cable inspection.

3.11 CONCLUDING REMARKS

TMB is now over 16 years old. As already mentioned, the paint system is starting to show its age, particularly on the main cable, and this will be attended to in the next few years. There have also been the problems of extrusion and wear of PTFE in the vertical bearings. The invention and development of the new sliding material was of obvious benefit in terms of not having to regularly jack up the bridge to change the PTFE disks. Like other bridges of similar age, there are plans to retrofit the main cable with a dehumidification system, as it is currently the only option available to bridge owners to limit or halt the loss of strength caused by wire breaks through corrosion.

REFERENCES

Beard, A.S., C.K. Lau, and E.H. Norrie. 1995. *The planning and implementation of the Lantau Fixed Crossing*. Proceedings of the International Bridge Conference: Bridges into the 21st Century. Hong Kong Institution of Engineers.

Beard, A.S., and J.S. Young. 1995. *Aspects of the design of the Tsing Ma Bridge.* Proceedings of the International Bridge Conference: Bridges into the 21st Century. Hong Kong Institution of Engineers.

Chan, W.M., R.J. Feast, J.D. Gibson et al. 2002. *Maintenance strategy of long span cable-supported bridges in HKSAR.* Proceedings of International Conference on Innovation and Sustainable Development of Civil Engineering in the 21st Century Beijing.

Charge, A., J.D. Gibson, and K.M. Hung. 2010. *Inspection and maintenance of long span bridges in Hong Kong.* Proceedings of the Seventh International Cable Supported Bridge Operators' Conference, Runyang Bridge, Jiangsu Province, China.

Chow, C.K., R. Hodges, C.K. Lau et al. 1982. *The planning and design of the Lantau Fixed Crossing.* Hong Kong Institution of Engineers.

Gibson, J.D. 2001. *Inspection and maintenance of Hong Kong's long span bridges.* Proceedings of Bridge Design, Construction and Maintenance Conference Institution of Civil Engineers Hong Kong.

Hunter, I.E. 1995. *Tsing Ma Bridge—Superstructure construction and engineering aspects.* Proceedings of the International Bridge Conference: Bridges into the 21st Century. Hong Kong Institution of Engineers.

Jennings, R.W. 1995. *Tsing Ma Bridge—Hong Kong excavation and caisson construction.* Proceedings of the International Bridge Conference: Bridges into the 21st Century. Hong Kong Institution of Engineers.

Law, P. 1995. *Tsing Ma Bridge—Construction of the towers and anchorages.* Proceedings of the International Bridge Conference: Bridges into the 21st Century. Hong Kong Institution of Engineers.

Phillips, D., and D. Sheen. 1995. *Modern and long span bridges from a railway perspective.* Proceedings of the International Bridge Conference: Bridges into the 21st Century. Hong Kong Institution of Engineers.

Proceedings of the International Bridge Conference: Bridges into the 21st Century. 1995. Hong Kong: Hong Kong Institution of Engineers.

Slater, C. 1999. *Tsing Ma Bridge.* Highways Department, Hong Kong Special Administrative Region.

Chapter 4

Storebælt East Suspension Bridge

Kim Agersø Nielsen, Leif Vincentsen,
and Finn Bormlund

CONTENTS

4.1 HISTORY

The 18 km long fixed link across the Storebælt Strait, often referred to as the Storebælt Fixed Link or the Great Belt Link, comprises two bridges and a tunnel. It is the only fixed link for road and rail traffic between east and west Denmark; see Figure 4.1. Construction of the link took place between 1989 and 1998. The political will was that the link should be built as two separate entities, one for rail and one for road traffic, and that the rail link should open before the road link.

The rail link opened in 1997 and consists of a low-level bridge across the western stretch of Storebælt and a bored tunnel below the eastern stretch. The two channels are separated by the small island of Sprogø. When construction of the fixed link was complete, the island had quadrupled in size as a result of the dredged material deposits. The road link opened in 1998 and consists of a low-level bridge across the western stretch, running parallel to the rail bridge and supported on common piers (the West Bridge) and a suspension bridge across the eastern stretch (the East Bridge).

This chapter deals only with the East Bridge, one of the world's longest suspension bridges, which has a main span of 1624 m (see Figure 4.2).

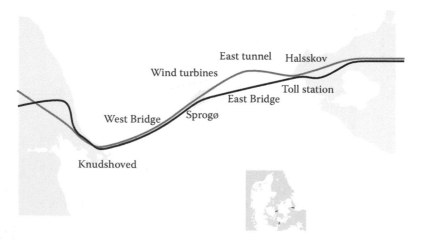

Figure 4.1 Alignment of the Storebælt Fixed Link.

Figure 4.2 The East Bridge.

The Storebælt Strait connects the Baltic Sea to the North Sea and handles most of the water flowing in and out of the Baltic. It is also the main shipping route to and from the Baltic Sea. Today, marine traffic through the Storebælt Strait totals about 25,000 vessels per year. These two facts have had a big influence on the design of the Storebælt Fixed Link and in particular on the design of the East Bridge. In the 1987 act for the fixed link, it was specified that the link should be designed in such a way that it would have no effect on the marine environment in the Baltic Sea, i.e., a "zero solution."

The fixed link replaced ferry services across the Storebælt Strait and has made crossings much easier, faster, and cheaper compared to the ferry era. Furthermore, emissions have been reduced by the move from the relatively heavy pollution of ferry and aircraft transport to rail and road.

In 2013, road traffic across the link totaled 30,000 vehicles a day, compared to 1997, when 8300 vehicles a day were transported by ferries; see Figure 4.3.

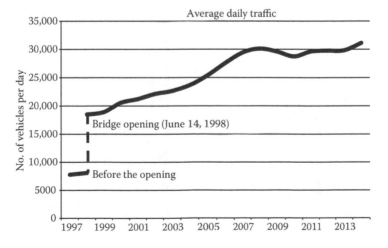

Figure 4.3 Development in road traffic.

The construction of the East Bridge took place from 1991 to June 14, 1998. The Storebælt Fixed Link is owned by the state-owned company Sund & Bælt Holding A/S (Sund & Bælt) through its subsidiary A/S Storebælt. Sund & Bælt is responsible for the operation and maintenance (O&M) of the Storebælt Fixed Link.

4.2 DESIGN

The 6.8 km long East Bridge, which carries a four-lane motorway plus emergency lanes, comprises two approach bridges (2518 and 1556 m) and a 2694 m long suspension bridge with a main span of 1624 m between the two concrete pylons and side spans of 535 m from anchor block to pylon. The navigational clearance is 65 m. The approach bridges have a span length of 193 m between the concrete piers.

The bridge girders are closed steel box girders with all stiffeners placed inside, resulting in a smooth outer surface. The inside of the box girders is protected from corrosion by a number of dehumidifiers.

The design of the East Bridge was the result of close cooperation between architects and engineers from the very beginning. The Danish architects, from Dissing+Weitling, together with a joint venture between two Danish engineering firms, COWI and Rambøll, were responsible for the much-admired outcome.

Two key elements underpinning the elegant profile of the 254 m high pylons were the elimination of the traditional crossbeam under the bridge deck and the unique shaping of the lower part of the pylons and the two crossbeams; see Figure 4.4. Additionally, the design of the anchor blocks as open structures, the light support of the girders on the slender bridge piers, and the detailing of crash barriers resulted in a very harmonious bridge

Figure 4.4 The elegant East Bridge.

structure. In addition, the abutments, which provide access to the interior of the bridge girders, are carefully integrated into the landscape to retain the bridge's elegant and airy appearance.

To warn aircraft and shipping—and to enhance the architectural expression of the bridge—lighting was installed on the pylons and anchor blocks and along the main cables. While the outer surfaces of the bridge structure are not illuminated, the anchor blocks are lit from within to emphasize the open and sculptural design.

The main cables have a diameter of 827 mm and consist of 18,648 parallel galvanized high-strength steel wires. The wires are protected against corrosion by a zinc paste and wrapped with a 3 mm thick galvanized steel wire further protected by three coats of paint. Every 24 m a pair of hangers support the bridge girder. The hanger cables are constructed from circular and z-shaped galvanized wires protected by polyethylene sheathing.

At the pylon top and at the anchor blocks, big cast steel saddles support the cables. The splay chambers in the anchor blocks are dehumidified as is the case with the box girders and for the pylon saddles. The dehumidification systems keep the relative humidity at 40%.

The slender bridge girder is continuous from anchor block to anchor block and, thus, passes without interruption between the pylon legs and again from anchor blocks to abutments. This reduces the number of expansion joints, which means that those at the anchor blocks need to absorb significant movement of up to ±1 m. Vertical bearings are fitted at the pylon legs in order to restrict the bridge girder against out-of-plane horizontal movements.

The design of the bridge girders was optimized in accordance with stress analysis and aerodynamic conditions. In order to protect the slender girders against oscillations due to vortex shedding from crosswinds, flaps were mounted on the underside of the main span girder. The steel girder of the suspension bridge is 31 m wide and only 4.34 m deep at the centerline of the box, which is streamlined with wind noses based on wind tunnel testing. The approach bridges have a girder depth of 7.10 m; see Figure 4.5.

In addition, the approach span girders are equipped with a total of 32 tuned mass dampers (TMDs) inside them in order to keep oscillations at an acceptable level. A TMD consists of a 7- to 8-ton steel plate suspended in a number of springs. Hydraulic buffers were installed at the anchor blocks in order to absorb traffic induced rapid movements from the suspension girder.

The road surface has a waterproofing membrane at the bottom and 55 mm mastic asphalt in two layers at the top. Road grip is ensured through crushed stones rolled into the surface.

The East Bridge extends over an international waterway to and from the Baltic Sea with water depths of up to 60 m. To minimize the risk to shipping, the main span of the East Bridge was determined on the basis of extensive ship maneuver simulations. Furthermore, the structures closest to the ship navigation route were designed so as to ensure that any collision from ships would not seriously damage the bridge. Risk studies and

Figure 4.5 Bridge girder section for approach span.

risk analysis formed an important basis for the design, construction, and operation of the infrastructure. These studies were behind the decision to establish and operate a vessel traffic surveillance (VTS) system to monitor shipping out to 12 nautical miles (nmi) on both sides of the bridges, based on radar, infrared cameras, and a very high-frequency radio system.

The basis for the risk analysis was established. Risk-acceptance criteria identified two main types of risks: traffic disruptions lasting more than 1 month and fatal accidents to users. The main risk of disruption for the East Bridge was found to be ship collisions. Other risks of a much lesser magnitude were found to be scour, earthquakes, road accidents, aircraft collisions, and discharges from ships.

4.3 CONSTRUCTION

Two methods of cable erection were prepared for the tender design: aerial spinning and prefabrication of strands. The tender prices, however, clearly indicated that the aerial spinning method was the most economically advantageous. The 254 m high pylons were tendered in both steel and concrete: the anchor blocks based on prefabrication or cast in situ and the approach girders as prefabricated concrete or steel solutions with different span lengths. The tender documents were divided into four packages to be priced by the contractors. The four packages were (1) superstructure and (2) substructure inclusive pylons for the suspension bridge and (3) superstructure and (4) substructure for the approach bridges.

In December 1990, the tenders were received from eight prequalified contractors. The tender evaluation was based on identification of the most economical solution. This included evaluation of costs of optional items, necessary upgrading of alternatives, differences in O&M costs, and differences in environmental implications and in the owner's risks. The tender evaluation for the substructures for the approach bridges as well as for the main bridge—based on prefabricated bridge piers, caisson anchor blocks, and concrete pylons—resulted in a contract with the Great Belt Consortium, which comprised Hochtief AG (Germany), Hollandsche Beton-en Waterbouw BV (the Netherlands), Wayss & Freytag AG (Germany), E. Pihl & Søn A/S (Denmark), and KKS Enterprise A/S (Denmark).

For the superstructure, the tender evaluation showed that the steel solution was the most economical with an approach span length of 193 m (increased from 168 m in the tender design) with the contract for both superstructures being awarded to the consortium of Costruzioni Metalliche Finsider Sud SpA (Italy) and Steinman, Boynton, Gronquist & Birdsall (United State). The construction contracts were signed on October 22, 1991, and the detailed design was developed by A/S Storebælt's consultants in 1991–1993 for substructures as well as superstructures.

The basis for the foundations of the concrete pylons, which were 25 m below sea level, consisted of placing prefabricated concrete caissons on compacted stone beds prepared under water at each pylon location. Due to the lower and more variable strength and deformation parameters of the clay below the stone beds, the thickness of the beds was increased from 1.5 to 5.0 m and, thus, placed on top of the stronger and more uniform marl strata. The prefabricated pylon caissons are 78×35 m² and divided into cells. After tow-out from dry dock, 30 nmi from the pylon location, they were filled with sand. The remaining parts of the pylons were cast in situ using climbing formwork 4 m in height. The anchor blocks were also prefabricated caissons, placed 15 m below sea level on wedge-shaped stone beds on top of a thick clay layer. The caissons for the anchor blocks have rectangular dimensions of 121.5×54.5 m². These caissons are also divided into cells and are ballasted with sand, olivine, and iron ore and the rear section was filled with concrete.

The concrete used for the substructure, including that for the pylons, mainly consisted of two types, a mix B for moderately exposed structures and a mix A for extremely exposed structures, e.g., in the splash zones. Both mixes included the use of low-alkali, sulfate-resistant Portland cement (A/HS/EA/G), fly ash, and microsilica, and crushed granite such as coarse aggregates and sea-dredged sand. The maximum water/cement ratios were 0.4 (type B) and 0.35 (type A) and the maximum water content was 140–135 L/m³. Extensive pretesting of concrete quality and workability was carried out, with protective precautions also being introduced in relation to evaporation, temperature control, and deformations during and after transport and casting of the concrete in order to fulfill the necessary long-term durability of the final structure. As this high-performance concrete with regard to durability had another performance regarding workability and precasting, education and training was planned and included in the contract.

The upper part of anchor blocks and the pylons were all in situ concrete works beginning at the top of the prefabricated caissons. The steel girders were produced over many distinct stages. The steel panels were fabricated near Livorno in Italy. They were based on Italian high-strength steel and transported to assembly yards in southern Italy and in Portugal. From here, the girder sections were transported to northern Denmark and welded together into bridge sections for transport and erection at the eventual bridge site at Storebælt. The steel types used and welded were Fe 510 D and Fe 420 KT TM (pr EN 10113). These materials were specified with a restricted carbon equivalent and the steel plates as delivered had an average of only about 0.33. These low values produced steel with high weldability, limited requirements for preheating, and reduced risk of hardened zones.

The exterior of the box girders had the following surface treatment:

- Blasting to Sa 3, using as reference the 1988 edition of International Organization for Standardization (ISO) standard ISO 8501-1
- Epoxy zinc primer (dry film 50 my)
- Epoxy polyamide intermediate coat (dry film 125 my)
- Aliphatic polyurethane topcoat (dry film 75 my)

Areas damaged during transport or from welding of sea fastening, etc., were generally repaired in a controlled atmosphere in smaller cabins or, if required, under large scaffolding and canvas covers built to suit the specific area. For the erection of the approach span girders at the bridge site, two sheer leg cranes were used, one floating and one fixed. The latter was built specifically for this work. The girders were placed on an elevated temporary support at their outer end. This was done in order to establish a bending moment in the previous girder during welding, which would be straightened afterward by lowering of the outer end.

The 57 suspension bridge sections (33 in main span and 12 in each side span) were equipped with four lifting points connected by two heavy bolts to a preinstalled stiffener arrangement in the girder section. During lifting, the lifting points were linked by a spreader beam. The bridge sections were lifted by two of four gantry cranes built for this purpose and placed on the main cables using a floating crane.

Spinning of the main cables was on the critical path of the working schedule because the start of the completion activities, such as compaction, wrapping, erection of clamps and hangers, and finally erection of girders, depended on the cables being completed. The method of spinning used was the following:

- Aerial spinning with controlled tension
- Spinning installations at only one anchor block
- Two spinning wheels on each tramway system
- Four loops on each wheel
- Alternate spinning on the two cables

The 5.38 mm cable wires were fabricated by Rylands–Whitecross Ltd. in the United Kingdom and by Redaelli Aps in Italy. To balance the horizontal forces on top of the pylon during cable spinning, the top saddles had to be offset from their final position. Due to the slenderness of the East Bridge pylons it was, however, possible to directly pull back the tops of the pylons toward the anchor blocks by 1.24 m using eight 48 mm diameter cables fixed at the top of the pylons and to the anchor blocks. The necessary horizontal force was calculated to be 8000 kN.

In all, the cable spinning work was completed in 137 working days. After spinning, the cables were compacted using four compaction machines. The average progress of compaction was 10 m/day/machine. The wrapping of the compacted cables was done using soft annealed steel wires and a two-component polyurethane zinc paste; see Figure 4.6.

The painting of the main cables was done inside eight heated painting shelters, each 110 m long, and after removal of a plastic film, which had been temporarily protecting the wrapped cables from salty air. The painting system consisted of an epoxy primer in two layers (minimum 40 my) with good adherence to galvanized wires and a polyurethane topcoat (minimum 110 my) with good elasticity and good durability when exposed to severe weather conditions and compatibility with the primer.

Road surfacing on the bridges included highly specialized work performed by several specialist contractors and inside 120–160 m long shelters, which were climate controlled for temperature and humidity. The steel surface was blasted to a degree of cleaning to Sa 3 according to ISO 8501-1, using vacuum grit-blasting equipment in general and open-air blasting along curb and base plates.

One hour following application of an adhesive, Casco Nobel 5000, by rollers (300 g/m²), the mastic was poured directly from dumpers onto the steel deck and leveled by men on their knees using steel board floaters and keeping the thickness within 4 ± 1 mm.

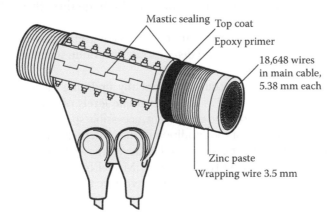

Figure 4.6 Layout of cable.

The 55 mm mastic asphalt layers (25 mm intermediate and 30 mm wearing course) were laid by three machine pavers inside 280 m shelters on the suspension bridge and in the open air at the approach spans. The bearings were free sliding, spherical types manufactured by Fip Industriale Spa in Italy. At the pylons, special bearings to bear the horizontal forces from the suspended structures were installed. They were sliding elastomeric bearings. The expansion joints were produced by the German company Maurer Söhne and have a maximum movement of up to 2000 mm for the biggest joints at the anchor blocks.

4.4 OPERATION

The road link across Storebælt opened to traffic on June 14, 1998. The total construction costs for the road and rail link amounted to DKK26.5 billion at current prices, with roughly 50% spent on each link. Because the financing was based on loans raised in the Danish and international capital markets—and guaranteed by the Danish state—the overall debt at the time of the opening totaled approximately DKK36 billion. The loans are serviced through revenue from the motorists using the link and from a fixed fee to the owner from Rail Net Denmark.

The Storebælt road link is the first user-paid motorway in Denmark with a toll station operating on the east side of the link and comprising 13 westbound and 11 eastbound lanes for automatic payment, which uses vehicle onboard units in fast lanes (30 km/h), or manual payment, which uses either credit cards in self-service lanes or cash and card payments in serviced lanes.

The focus areas for Sund & Bælt's O&M of the fixed link are to

- Meet customer expectations with regard to traffic safety, accessibility, and convenience;
- Protect and enhance the company's infrastructure facilities;
- Ensure optimum administration and high quality throughout the company's activities; and
- Engage in proactive measures in relation to environmental impact, health and safety, traffic safety and demonstrate corporate social responsibility.

Meeting these objectives requires continually updating of documentation, procedures, and plans as well as the appropriate levels of organization. Also important are effective tools and equipment, accessible spare parts, and continually updated knowledge of the condition of the structures and installations.

The overall management system used as the framework for the company's O&M activities is based on the European Model for Business Excellence. Specific operational objectives are defined every year as a supplement to the strategic objectives within the four result areas within the model.

The reliability of the Storebælt Link is a key issue because it is critical for mobility between east and west Denmark. The traffic safety is based on continuous risk evaluations, early warnings for drivers on changing road, and weather and traffic conditions, as well as safety measures for drivers and workers during maintenance work. Furthermore, the O&M activities are planned so that the effect on traffic is minimized, e.g., by using platforms for easy access but which do not disturb the traffic and performing activities on the road during times of low traffic intensity.

To achieve low vulnerability and long durability and to keep the visual impression of well-maintained structures, the O&M work must be carried out in a systematic and efficient way. The procedures and systems to support this are based on the Maximo maintenance management system and the Meridian document-handling system. Another prerequisite for efficient O&M work is a competent organization. Sund & Bælt, the owner and operator of the fixed Storebælt link, is a small organization staffed mainly with managers tasked with the planning, managing, monitoring, and optimizing of all activities. Most activities are outsourced to specialists at contractors and consultants for periods of typically 3–5 years. Consequently, all knowledge on the actual conditions of the infrastructure and about Sund & Bælt's customers (users of the link) must be kept within the organization. The experience from the annual inspections and maintenance work plans for inspections and evaluation of conditions means that periodic preventive maintenance and service activities are reevaluated in order to optimize resources and outcome and also plan for possible reinvestment and the purchase of critical spare parts. Specially designed platforms are used to perform the inspection and maintenance work efficiently and without disturbance to traffic; see, for example, Figure 4.7. Inspections that involve absciling have also been performed; see Figure 4.8, e.g., for inspection of concrete surfaces of the pylons.

Figure 4.7 Platform underneath the bridge girder.

Figure 4.8 Concrete inspection performed by abseiling.

4.5 INSPECTION

The technical O&M of the Storebælt Fixed Link is based on guidelines in an O&M manual plus supplementary procedures and instructions. This is part of Storebælt's overall integrated management system for quality, environment, and safety.

The maintenance management system used on Storebælt is based on a standard program, Maximo, which has been adapted for Storebælt's use. Maximo is a web-based program to be used from a personal computer (PC) with an Internet browser. It contains all the information necessary for planning, execution, and reporting of O&M activities (time, resources, and technical information). Results from the activities are recorded in Maximo and are used for optimization of the O&M work.

The inspections of steel and concrete structures are divided into three types: routine, principal, and special inspections.

4.5.1 Routine inspection

Routine inspections are carried out as a follow-up between principal inspections. The purpose is to ensure that the structure continues to function correctly in order that road safety and the desired condition of the bridge are maintained and also to follow up whether there are any special maintenance issues relating to the inspected sections. The result of the inspection means that an assessment can be made as to whether there is a need to change its frequency. The usual period between routine inspections varies from 6 months to 1 year. Routine inspections are normally carried out by a consultant or a Storebælt engineer.

4.5.2 Principal inspection

The purpose of a principal inspection is to monitor the condition of all elements of the structure in a systematic way. This ensures that maintenance work can be planned and conducted in a technically and economically optimal way, a timely manner, and in an optimal sequence. A principal inspection is a systematic examination of the status of all components of the construction element, e.g., expansion joint. The inspection must record the element's condition and any damage and whether there is the need for a special inspection, maintenance, or replacement of parts.

The report for a principal inspection will also specify the condition and residual life of the components and an assessment of the adequacy of the procedures/instructions for preventive maintenance. The normal period between principal inspections varies from 3 to 6 years.

General inspections are performed by a qualified person, typically a Storebælt or consulting engineer.

4.5.3 Special inspection

Special inspections are performed either after an already completed inspection plan, as recommended by the routine or general inspection, or when a structural element has been exposed to extreme stresses. The aim of the inspection is to provide a more accurate report on the condition of the object, e.g., with regard to the extent of damage and root cause, and to decide on repair strategies or, where possible, acceptance of the condition. The special inspection is normally performed by a qualified person and possibly with the assistance of a specialist company or a specialist in that field.

4.6 EVALUATION

Apart from the normal routine inspections, carried out over the 15 years of operation, some elements of the East Bridge have been the subject of further investigation or evaluation. The evaluation of the elements is typically initiated by observations carried out at a routine inspection, deviations from the design basis, special observations, or similar. Examples of evaluations or investigations carried out since the opening in 1998 are set out below.

4.6.1 Guide vanes

During the final phases of girder erection and road-surfacing works, the suspended spans of the Storebælt East Bridge displayed substantial wind-induced oscillations, which were deemed to be unacceptable to future users of the bridge. The single-mode harmonic nature of the oscillations, together with the correlation with similar observations during wind tunnel testing, suggested that vortex shedding was responsible for the excitation. With the aid of wind tunnel section model testing, a guide vane design was devised for mitigation of the undesirable oscillations.

Manufacture and mounting of the guide vanes, along with disassembly and removal of temporary installations, was completed on June 12, 1998, 2 days before the grand opening ceremony; see Figures 4.9 and 4.10. The structural monitoring system was retained in the hope of obtaining evidence of the efficiency of the guide vanes, but theft of the data-acquisition computer put an unexpected end to it. This was unfortunate because the weather kept challenging the bridge with 5–12 m/s northerly or southerly winds throughout June 1998—ideal conditions for vortex shedding excitation. Frequent visual inspections by the bridge staff revealed no sign of girder oscillations. Regular inspections of the bridge in northerly or southerly winds were continued, now aided by a video camera. After 9 months of service the reports remained the same—no vortex-induced oscillations. In conclusion, guide vanes may prove very efficient for suppression of vortex shedding excitation of shallow box girders provided that they are positioned correctly.

Figure 4.9 Guide vanes in main span.

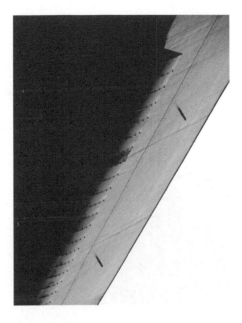

Figure 4.10 Guide vanes.

4.6.2 Vibration of hangers

The East Bridge has handled wind speeds of up to 20–30 m/s on several occasions without any significant movement of the hangers. However, in March 2001 extreme hanger movements (amplitude 1 m) were observed at a wind speed of 15 m/s in combination with wet snow.

In order to prevent such movements, various damping systems were tested:

- Wind ropes: horizontal 16 mm wires connecting the 12 longest pairs of hangers; the wires were prestressed to 35 kN
- Spiral ropes: steel wire in a 16 mm tube wound around the hangers with a pitch of 1.0 m
- Tuned liquid dampers: a glass fiber box with 10–20 cells partly filled with a mixture of glycerin and water

Following 4 years of tests with no conclusion on the optimal damper system, it was decided to install and test hydraulic dampers that had been successfully used on many other bridges, especially cable-stayed bridges; see Figure 4.11. The effectiveness of hydraulic dampers is independent of the cable vibration mechanism because they simply increase the damping ratio of the hanger.

Figure 4.11 Hydraulic damper system.

The hydraulic dampers are expensive and more difficult to implement on an existing bridge compared to the other damping measures tested so far. Two hydraulic damper systems were installed at two hanger locations in 2005, both of them on the fatigue-critical second hanger from the pylon. The system demonstrated its effectiveness and is now installed on the 16 hangers closest to the pylons. A system has been installed to monitor the performance of the dampers and predict the fatigue lifetime of the hangers. Two accelerometers are mounted approximately 4.5 m above the bridge deck at each of seven hangers. Two of the monitored hangers have no damping. The data from the accelerometers indicate an acceptable lifetime for damped and nondamped hangers at the East Bridge.

4.6.3 Safety barriers

Significant vibrations of the outer safety barrier handrails were observed shortly after the East Bridge was opened to traffic. The vibrations were often seen at transversal wind speeds of approximately 10 m/s. Strain gauge measurements were therefore carried out on the post top plate and at cut-outs in the handrails. The strain gauge measurements showed stresses varying between 75 and 100 MPa in the top plate. The following tests for damping the vibrations of the handrail were carried out in order to avoid fatigue problems in the top plate:

Figure 4.12 Neoprene blocks between handrail and the internal prestressed wire.

- Stock bridge dampers on the handrails tuned to the handrail frequencies
- Straps between handrail and guardrail
- Closing the opening on the underside of the handrail
- Neoprene blocks between the handrail and the internal prestressed wire; see Figure 4.12

After a testing period of about 6 months, it was decided that neoprene blocks fitted in the handrail provided the best and most economical solution. Neoprene blocks are now installed inside the handrails at every 4 m across the 6.8 km East Bridge.

4.6.4 Concrete

In 1987, an expert group was established to develop the basic concrete specifications. The requirement was that the concrete should have a service life of 100 years. With the existing guidelines, there would probably not be any significant deterioration due to frost or alkali silica within 100 years, but it was likely that they would not be able to prevent reinforcement corrosion from chloride ingress.

In preparing the specification requirements of the Storebælt concrete, special attention was therefore paid to its ability to stop chloride ingress. In contrast to frost and alkali silica damages, chloride ingress can only be slowed and not countered completely by modifying the concrete specifications. Before this, concrete had not been designed specifically to resist chloride ingress. Thus, the specification for the Storebælt concrete was both to achieve 100 years of service life regarding corrosion due to chloride ingress and to develop tools to calculate chloride ingress into concrete.

In order to achieve a more dense concrete, two significant changes were introduced to the Storebælt concrete compared to that previously used, namely, a very low water/cement ratio (about 0.35) and addition of both silica fume and fly ash. New methods were developed and tested to design the concrete's resistance to chloride ingress. A study of the concrete's actual durability was conducted in 2009 as part of an overall summary of results for chloride-induced corrosion in noncracked concrete on the Storebælt structures. The purpose was to summarize the results and experiences from concrete investigations after 17 years of operation. The report was limited to noncracked concrete and was based on studies of selected areas of Storebælt. In addition to achieving the desired durability in the toughest exposed areas, the studies revealed that a service life of significantly longer than 100 years was achieved in noncracked concrete in all other areas of the structures. The toughest exposed areas apparently include only a local area around 0–2 m above sea level. With a relatively low-maintenance effort, the life span of the Storebælt concrete structures could be extended for hundreds of years.

4.6.5 Fatigue evaluation for orthotropic steel bridge deck structure

Since its opening in 1998, the traffic growth on the Storebælt Link has been significantly greater than forecast. In 1998 the average daily traffic was 18,500 vehicles and this number has increased to more than 30,000 today. This extraordinary traffic growth, together with the increased number of heavy goods vehicles—some up to more than 120 tons—has necessitated an evaluation of the fatigue capacity of the orthotropic steel deck.

The composite action between the steel deck and the pavement has a favorable influence on the fatigue life. This effect, together with the influence of the temperature of the pavement, has been investigated as part of the evaluation that was carried out. The results of the field test were compared to the original design assumptions. A particular focus was on the transverse butt welds of the trapezoidal stiffeners and longitudinal trough-to-deck plate welds. The reassessment was based on revised traffic forecasts as well as on load tests and monitoring programs, which had been carried out between 2000 and 2007.

A fatigue-monitoring system has been installed within the bridge girder. The system comprises a central data logger and 15 uniaxial foil strain gauges attached to the deck plate structure at longitudinal trough-to-deck plate welds, as well as at transversal welds in trough joints between deck sections.

The evaluation of the fatigue capacity of the orthotropic steel deck concluded that the bridge deck will have sufficient fatigue life to reach the planned design life span of 100 years. This is despite the fact that traffic growth has been considerably greater than foreseen in the initial forecasts.

Furthermore, the evaluation concluded that the maximum fatigue damage occurs at butt-welds in trapezoidal stiffeners. It can be concluded that in case of fatigue damages occurring in the deck structure, this would most probably happen in the transversal butt welds of the trapezoidal stiffeners, which are simple to test and repair—compared to longitudinal welds, for example.

4.7 MAINTENANCE

The majority of maintenance work on the East Bridge has so far dealt with moving parts and surface treatment, e.g., replacement of rubber components in the expansion joints and repainting at the bridge girder and bearings. In addition, some specified work, such as tightening of cable clamps or repair of issues arising from construction, has been carried out. Some examples of maintenance work undertaken are set out below.

4.7.1 Cable clamp rods

The cable clamps are fastened at the main cables with rods $\varnothing$ 44 mm (thread M48) and the nominal tension in the rods is 1 MN. The specifications require a retension of the rods if the tension falls below 80% of the nominal tension. During construction, all the screwed rods were tensioned four times. Tension loss was monitored by micrometer measurements on 40 selected inspection rods, timed with successive increases in tension in the main cable due to erection works and surfacing. Micrometer measurements during the construction phase were carried out from the catwalk, but this turned out to be very difficult to perform when operating from the top of the main cable.

The measurements were carried out from the cable gantry (see Figure 4.13), and in 2004, the tension in the bolts was around 80% of nominal tension. This called for a new retension campaign from 2004 to 2007. The retension was carried out using the original hydraulic equipment. About 20 special load cell washers were placed between the nuts and clamps to simplify the future measurements. The load cell washers could be read by handheld equipment from the main cable. The measurements were accurate and easy to perform, but the load cells started to fail after approximately 5 years. The sixth retension of the rods started in 2013 and was based on tension measurements made by hydraulic jacks.

Measurements made by hydraulic jacks require access from the cable gantry. This is time consuming and expensive. Bolt tension can be measured with special ultrasonic equipment and laboratory tests done in 2013 showed the necessary accuracy. The technique requires smooth and parallel rod ends and 20 special rods mounted on different clamps. The ultrasonic measurements can be carried out from the top of the main cable and will

Figure 4.13 Cable gantry.

be used to decide on the time when the seventh retension of the rods will be required.

4.7.2 Anchor blocks and concrete surfaces

During construction of the anchor blocks, significant crack formation appeared on the top deck (foundation deck) of both anchor blocks; see Figure 4.14. The crack formation consisted of cracks with widths of up to 0.5 mm. The cracks appeared in form of coarse net cracking with mesh size varying from 0.4 to 4 m. The top deck had been temporarily repaired with acrylic modified grout of up to 30 mm in thickness on several occasions. The condition of the grout was very poor with defects such as cracking and spalling; adherence was also poor. Crack injection with polyurethane was also carried out. There were many minor holes, irregularities, and depressions in the concrete surface, and water gathered in many of the depressions after a wet weather. The slope conditions were not effective due to these irregularities and depressions.

In 2001, a pilot project was carried out to verify which repair method should be used for the restoration of the top deck. The project was to verify the applicability of the repair method and the practicability of the project, as well as transportation of materials and equipment to and from the anchor blocks. The focus was mainly on testing several coating materials' ability to span over cracks and absorb movement.

The repair method was selected on the basis of the 2004 pilot project. The restoration work consisted of concrete milling, fixing of the stainless steel

Figure 4.14 Anchor block top deck before restoration.

anchors and stainless steel mesh reinforcement, casting of a new concrete layer (70–150 mm) with cement-based mortar with synthetic admixtures, and ending with a three-component acrylic flooring system. The project on both anchor blocks was completed in 2005. In 2007, the acrylic flooring system was installed on both pylons (top deck and both crossbeam decks).

4.7.3 Repair of painted surfaces

The paint system used at the bridge girder consists of the following:

- Zinc epoxy primer, 50 µm
- Epoxy polyamide, 125 µm
- Polyurethane, 75 µm

The overall condition of paint at the box girder is in good condition after 16 years. The surfaces exposed to sunlight are, of course, partly faded. The paint repair undertaken at the box girder was limited to section joints and minor areas subject to paint/paint repair carried out on site during the construction phase.

The paint system for the bridge bearings was the same as for the bridge girder. Bearings, which can be sprayed by seawater in stormy weather, have been subject to more intensive paint repair. The objective now is to investigate methods by which the bearings can be enclosed.

It was apparent that the guide vanes mounted underneath the girder along the main span had undergone insufficient surface treatment. They have been subject to intensive paint repair. All paint repair on the box girder, bearings, and guide vanes can be done from the platforms at the East Bridge without disruption to traffic.

4.8 BEST PRACTICES

4.8.1 Optimal management and appropriate performance quality

Prerequisites for efficient maintenance on the Storebælt Link are as follows:

1. Complete documentation from construction updated during operation
2. Clear procedures and instructions for the work updated on a continuous basis and based on experience in use
3. Competent employees, consultants, contractors, and suppliers
4. Updated maintenance plans, with the frequency depending on condition, functionality, criticality, etc.
5. Efficient maintenance systems for collecting and analyzing data from monitoring, supervision, maintenance, and operation
6. Accessible spare parts for vital or critical components with long delivery time or where there are major consequences to breakdowns (risk analyses and vulnerability evaluations are parts of an efficient management system)
7. Efficient tools and equipment and easy access to infrastructure

4.8.1.1 Updated documentation

It is very important that all drawings, procedures, and instructions are updated and reflect the changes made to the infrastructure during operation, as well as the experience gained from use of procedures and instructions. The routines must ensure that all documents are reviewed and are updated once a year by the responsible operations manager if necessary. As a result, it is important that there is input from service and maintenance contractors and from inspectors after completion of work.

4.8.1.2 Competent organization

A high degree of competence and strategic overview is acquired by the employees, in part through on-the-job training and learning from more experienced colleagues and business partners. However, it is also gained through planned personal development, together with training at internal and external workshops and conferences and established networks. Sund & Bælt has established international networks with other large infrastructure operators in order to exchange experience and investigate possible solutions together. Further benchmarking processes are performed with other organizations in order to learn the best practices from each other and to increase efficiency. The most recent benchmarking took place together with the operator of the Øresund Fixed Link between Denmark and Sweden in 2009.

4.8.1.3 Updated maintenance plans

In order to optimize resources and outcome, the plans for inspections, condition evaluations, periodic preventive maintenance, and service activities will be reevaluated on the basis of the experience gained from inspections and maintenance work. This takes place after recommendations from those involved, at least once a year and always when the plans are used for the next tender process.

4.8.1.4 Efficient maintenance systems

The maintenance management system used in Sund & Bælt is based on a standard program, Maximo, adjusted for Sund & Bælt use. It is a web-based program to be used from a PC with an Internet browser. It contains all information necessary for planning, execution, and reporting of O&M activities (time, resources, and technical information). Results from the activities are recorded in Maximo and are used for optimization of the O&M work.

4.8.1.5 Accessible spare parts

The need for the purchase of critical spare parts is defined on the basis of a systematic evaluation of the risks and vulnerability of the different parts of the infrastructure and their criticality for the use at the infrastructure. The parameters determining the stock of spare parts at the infrastructure include accessibility of suppliers, the uniqueness of the spare part, and the delivery time.

4.8.1.6 Efficient tools and equipment

Since all inspection and maintenance takes place on an infrastructure with traffic, it is important to use efficient tools for easy access and with as little disruption to traffic as possible; see Figure 4.15. Specially designed inspection platforms are used for the inspection of all surfaces on the East Bridge. The purpose is to minimize the time spent on the infrastructure for O&M, to improve the quality of work and minimize overall costs.

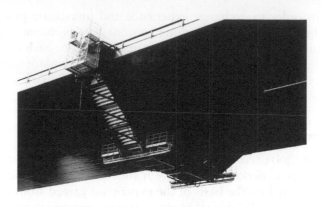

Figure 4.15 Maintenance platform.

4.8.2 Proactivity regarding environment, health and traffic safety, and social responsibility

Proactivity is determined to a high degree by awareness of the topic among employees and their knowledge of the infrastructure. It is also dictated by risk evaluations, planning that also focus on all the "soft" parameters, and, of course, the definition of conditions in contracts with suppliers, consultants, and contractors in these fields, as well as the follow-up on their satisfactory fulfillment through supervision and audits.

The management systems in the company relating to quality, environment, health and safety, traffic safety, and social responsibility form the basis for the awareness. Evaluation of risks related to the actual activity is essential for setting the requirement at the right level and with the correct focus. Paradigms are formulated for typical situations but should always be reassessed for each actual case.

Audit plans and plans for supervision are also based on evaluation of risks related to the activities in order to focus on the most hazardous areas. If incidents or near misses occur, they are always followed by evaluations as to whether corrective actions are needed in order to avoid reoccurrence or accidents. Emergency plans are prepared for the most hazardous situations in order to minimize the consequences of accidents.

4.8.3 Accessibility of the road link

The availability figure for the Storebælt road link is 99.8% (valid for closure for one or both directions) over the 15 years of operation. Complete closure of the road link primarily occurs in case of strong winds, traffic accidents, falling ice, and VTS alarm (for potential ship collision); see Figure 4.16.

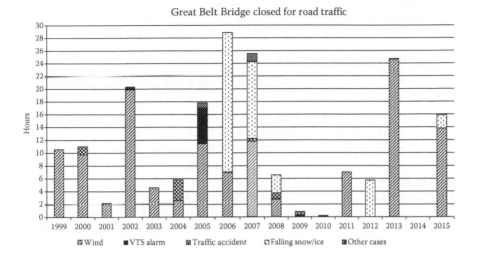

Great Belt Bridge closed for road traffic

Legend: ▨ Wind   ■ VTS alarm   ▥ Traffic accident   ▱ Falling snow/ice   ◪ Other cases

Figure 4.16 Closure of the Storebælt Link.

Since its opening in 1998, the Storebælt Link has been closed for about 90 h due to strong winds and about 42 h due to risk of falling ice. Complete closure of the link due to traffic accidents is relatively rare (about 7 h).

4.8.4 Impact on traffic from roadworks

Roadworks have an impact on traffic, of course, and drivers experience some degree of inconvenience when passing through or by them. One of Sund & Bælt's four core values is to be consumer focused and that means minimizing the inconvenience to drivers.

To minimize inconvenience to traffic, it is desirable to do the following:

- Draw up a good traffic management plan for the roadworks.
- Do other jobs during the same roadworks.
- Minimize the duration of the roadworks.
- Carry out the roadworks during periods of low traffic.

Only one contractor is allowed to set up traffic signage for roadworks at the Storebælt Link. This ensures a good and consistent quality of the signage and that personnel are familiar with the unique weather conditions on the bridge. The contractor has to comply with the procedure for roadworks at the Storebælt Link. The procedure describes how to plan roadworks and includes an appendix of approximately 20 approved traffic management plans. The procedure also describes under which conditions roadworks can be carried out at the link. For example, it is not permitted to undertake roadworks in either the slow or the fast lane when the traffic density is higher than 1200 vehicles/h or if the visibility is lower than 400 m.

4.8.5 Inconvenience index

Sund & Bælt has introduced an index for measuring the theoretical inconvenience to drivers passing through or by roadworks. The index makes it possible to compare and evaluate the inconvenience to traffic of completed roadworks and to estimate the impact of future roadworks on traffic. The index depends on the actual number of vehicles passing the roadworks and their locations. The number of vehicles passing the roadworks is multiplied by a factor, which depends on the number of vehicles and their locations. The factors are represented in Tables 4.1 and 4.2.

Every month the actual roadworks are evaluated. The time and duration of every single incidence of roadworks are compared to the actual number of vehicles and the index is calculated. Figure 4.17 presents the indices in 2010, 2011, 2012, and 2013.

The index is highest in spring and autumn. During winter, the traffic activity on the road is low, while in June and July major roadworks are avoided due to the high season for holiday traffic.

If the index is higher than 10, the procedure prescribes a further analysis of the cause. In the period from 2010 to 2012, the index was higher than 10 in April 2010, September 2010, and May 2012. In all three cases this was due to renewal of pavement.

When major roadworks are planned, an estimate for the index can be calculated based on the traffic forecast and expected traffic management plans. If the forecast predicts an index higher than 10, the planning of the work must be reassessed. If an index higher than 10 cannot be avoided, the drivers should be informed of the inconvenience that will be caused by the roadworks in advance.

Table 4.1 Factors depending on number of vehicles

No. of vehicles (x)	Factor
x < 600	1
600 < x < 1200	2
1200 < x	4

Table 4.2 Factors depending on location of roadworks

Location of roadworks	Factor
Emergency lane	0.25
Slow lane	5
Fast lane	3
All lanes, one direction	10

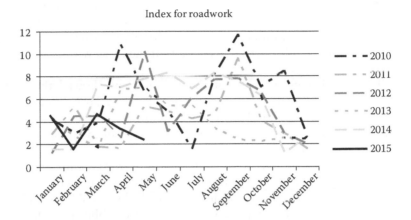

Factor:

Number of cars: > 600 ← 1200: current traffic figures ÷ 1200 × 2
                   > 1200 × 4

Roadway: hard shoulder × 0.25
                   right track × 5
                   left track × 3
                   1/2 bridge × 10

Figure 4.17 Indices for roadworks 2010–May, 2015.

4.9 FUTURE PLANS

Five- and 50-year plans are made to manage future reinvestments. The 5-year reinvestment plan is revised every year and the 50-year plan every fifth year. This gives the operations managers a good overview with regard to planning future projects and to allocating resources over the coming years. Three major projects planned for the near future are set out below.

4.9.1 Dehumidification of main cables

Inspection of the main cables, in 2008, revealed damage to the paint system. The upper half of the cable had some local areas where the topcoat had vanished, while some other areas had cracks in the paint. The damages to the paint system called for a renewal of the surface protection. At this point, the choice was between a repaint or wrap and dehumidification of the cable system. The two alternatives were compared in a cost–benefit analysis:

- Option 1: A complete repainting of the cable system in 2014 then wrapping and dehumidification of the cable system approximately 15 years later. This option included an internal inspection of the cable wires at the time of dehumidification.

- Option 2: Wrapping and dehumidification of the cable system in 2014. The lifetime of the system was estimated to be approximately 40 years, with minor service/inspection costs each year. This option did not include internal inspection of the cable wires.

The cost–benefit analyses of the two options showed the cost to be approximately equal. Option 2 was expected to give a better certainty of corrosion protection of the cable system from a long-term perspective. Therefore, the conclusion of the analyses was to install a dehumidification system at the two main cables.

Airflow investigations were made to minimize the number of air inlets and outlets for the dehumidification system. The tests were carried out by air injection at one spot with detection of leaks at cable clamps at a distance from the inlet. Air leaks were detected by bubble test. Three levels of air pressure were tested, 2000, 2500, and 3000 Pa. The bubble test of the clamps was made from the top of the cable and by rope access on the sides and at the bottom. Tests around the cable clamps were done by spraying soap and water onto the cable clamp sealing. If it was not possible to detect air at the sealing, the water drain holes under the clamp were used instead. The test results showed that it was possible to locate air leakage at a distance of up to 390 m away from the injection sleeve. See Table 4.3.

The sealing between the upper and lower cable clamps was tested to determine whether the sealing product needed to be replaced or partly renewed when wrapping the main cable and painting the clamps. The original sealing is from 1998 and built up in one layer. The total depth of the sealing was 102 mm, on the side of the cable clamps. The product was Starmastic P95 single-component polyurethane with a shore A45±5. After installation, the sealing was painted with the same topcoat as the cable clamp. Inspection of the sealing found cracks in the paint and also smaller cracks in the surface of the sealing; however, the cracks were only 1.5 mm deep. To undertake a shore hardness test of the sealing, four samples were cut out of random selected clamps. The test showed a shore value of the four samples at A43±3 and a good consistent flexibility according to ultimate-breaking-strength test.

In the light of the above tests and the information from the airflow tests, it was concluded that the sealing in the clamps was in a good condition. The minor leakages in the sealing have no significance for the operation of

Table 4.3 Test results

Pressure (P)	Airflow (m³/h)		Length of airflow (m)	
	Total	Each direction	West	East
2000	63	31	310	320
2500	72	36	335	345
3000	82	41	385	390

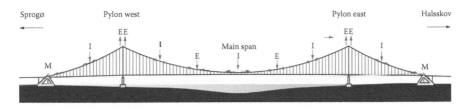

Figure 4.18 The final layout of the dehumidification system. E: outlet; I: inlet.

the dehumidification system. Therefore, the sealing in the clamps will not be renewed, just repainted with a layer of topcoat. The final layout of the tendered dehumidification system for the main cables at the Storebælt East Bridge is shown in Figure 4.18.

The project started in spring 2014. Wrapping of the main cables is managed from four cable crawler gantries, each 33 m long. The work on wrapping the cables and the installation of the dehumidification system inside the bridge girder is planned to be finished in early 2015.

4.9.2 Replacement of polytetrafluoroethylene sliding material in bearings

The bearings for the East Bridge were designed and manufactured by Fip Industriale Spa in Italy. In total, there are 50 vertical bearings on the approach spans and 4 on the suspension bridge; see Figure 4.19. At the pylons there are 4 horizontal bearings. The vertical bearings are freely sliding spherical bearings and the sliding surface is made of 6 mm PTFE and slides on a stainless steel plate. The upper 3 mm is placed in a recess and the remaining 3 mm is wearing surface.

Figure 4.19 Vertical bearing.

Inspections in 2013 have revealed differences in the wear. On the four bearings nearest the anchor blocks, the wear is down to a level that triggers replacement in 2014/2015. Plans for jacking up the bridge girder and dismantling the four bearings for service on the PTFE wear plate are ongoing. Similar work is expected to be done on the remaining 46 bearings within the next 7 years according to inspection results made for every fifth year.

4.9.3 New mastic asphalt for use as surfacing

The existing mastic asphalt surfacing at the East Bridge has generally displayed high performance since it was laid in 1998. However, the surfacing now shows some rutting in the heavy lane and the resistance to rutting will need to be improved when resurfacing is called for in future.

For safety reasons, rut repair to restore the transversal profile was conducted in 2011 using micro–asphalt surfacing. However, the use of micro–asphalt surfacing as remedial action is viewed as an interim solution to extend the service life of the existing surfacing. It is anticipated that a permanent solution involving resurfacing of the wearing course in the heavy lane will be done before 2020. To prepare for this resurfacing need, it was decided to run a laboratory investigation with the objective of developing a new mastic asphalt mix design for the wearing course.

Project objectives were defined with regard to the two main deterioration factors, i.e., rutting and cracks, using the history of the performance of the existing surfacing together with forecast future developments in the traffic load on the bridge:

- Objective 1: The resistance to rutting should be improved by 100% as compared to the existing wearing course.
- Objective 2: The resistance to development of cracks must not be decreased as compared to the existing wearing course.

Since the performance of the existing surfacing—with the exception of resistance to rutting—has shown excellent performance, it was decided to stay with the original mix design used and to look into the possibilities of using an elastomeric modified bituminous binder to improve the resistance to rutting. Five different types of bituminous binders, all recommended for use in bridge surfacing, have been tested. The laboratory program utilized the standard Danish methods for mix design supplemented by functional testing: a European test method (dynamic creep) to look into resistance to rutting and a Danish test method (three-point bending beam) to evaluate the resistance to cracking.

To evaluate the results from the functional testing, threshold values with regard to rutting and cracks were derived from the project objectives listed above. The threshold values were determined using testing slabs obtained

Table 4.4 Functional testing values

	Threshold value	Nynas Endura N5
Ball indentation[a]	–	33 mm
Ultimate penetration strength[b]	–	3.3 MPa
Dynamic creep[c] at 40°C	<1.4	1.3
Dynamic creep[c] at 50°C	<3.1	2.5
Three-point bending at 5°C	<944 MPa	758 MPa
Three-point bending at −10°C	<2175 MPa	1861 MPa

[a] Desired value between 30 and 34 mm.
[b] Desired value between 2.5 and 3.5 MPa.
[c] Creep rate in microstrain per pulse between 3000 and 3600 pulses.

from the quality control measures undertaken during the production of the existing mastic asphalt wearing course in 1998.

The findings of this investigation showed that the use of Nynas Endura N5 as a new binder in a mix design comprising the same aggregates as used in the existing mastic asphalt has the potential to improve the performance of the mastic asphalt surfacing as set out in the project objectives and demonstrated by the functional testing as shown in Table 4.4.

4.10 CONCLUDING REMARKS

The Storebælt Fixed Link has now been in operation since 1997/1998 and the structures are still in the early years of the planned design lifetime of at least 100 years. Until now no major reinvestments for the East Bridge have been made; however, they will start in the near future. The following four focus areas for O&M have proved their worth:

- Meet customer expectations with regard to traffic safety, accessibility, and convenience.
- Protect and enhance the company's infrastructure facilities.
- Ensure optimum administration and high quality throughout the company's activities.
- Engage in proactive measures in relation to environmental impact, health and safety, and traffic safety and demonstrate corporate social responsibility.

Some work has been done on benchmarking against other large cable-supported bridges. This is very complicated since they are different types of bridges, are of different ages, and are located in different countries. The International Cable Supported Bridge Operators' Association is an important forum for exchange of experiences. The best and most productive form of benchmarking is through discussion and comparison of different strategies between the different international operators.

Chapter 5

Forth Road Bridge

Barry Colford

CONTENTS

5.1 INTRODUCTION

5.1.1 Background and location

The Forth Road Bridge (FRB) (Figures 5.1 and 5.2) is a long-span suspension bridge and was opened in September 1964. At the time of opening, the bridge had the fourth longest main span (1006 m) in the world. The bridges that had longer spans than FRB were Golden Gate, Mackinac Straits, and George Washington. All these three bridges are in the United States, so the FRB was the first long-span bridge to be built outside the United States.

Figure 5.1 View looking northeast form the south side.

Figure 5.2 View looking southeast from Fife on the north side.

The bridge crosses the Firth of Forth about 15 km west of Edinburgh and is a vital link in Scotland's strategic road network. The bridge deck supports a dual two-lane carriageway without hard shoulders or strips. There is a separate footway/cycle track on either side. Over 25 million vehicles now cross the bridge each year. The historic importance of the structure to Scotland was recognized in 2001 when the bridge was classified as a category A–listed structure.

The bridge had been a tolled crossing since opening. However, in February 2008, tolls were removed on the FRB after an act of the Scottish parliament that removed tolls from all of Scotland's tolled bridges. Funding for maintenance and operation of the bridge now comes from a Scottish government annual grant.

5.1.2 Details of the construction

The bridge has a main span of 1006 m and the side spans are each 408 m long. The orthotropic deck on the main span is made up of a 12.7 mm stiffened steel plate overlain with 38 mm thick mastic asphalt, on a waterproofing layer. The deck to the side spans is a 203 mm thick reinforced concrete slab with a similar surfacing detail to the main span. The decks on both the main and side spans are supported on steel stringer beams that span between large steel cross girders spaced at 9.14 m center to center. These cross girders are supported by two longitudinal stiffening trusses, which, in turn, are supported by the main cables. The main cables are 600 mm in nominal diameter and are made up of 11,618 of 4.98 mm diameter galvanized wires that transfer the loads from the bridge to the main and side towers and also down to the north and south anchorages. The anchorages are tapered rock tunnels, filled with concrete and posttensioned. Linking the stiffening trusses to the main cables are 192 sets of steel wire rope hangers at 18.29 m centers that vary in length from 2.5 m at midspan to 90 m adjacent to the main towers.

The cables at the main towers, at the side towers, and in the anchorage chambers are seated directly on cast saddles. These saddles on the main tower are fixed directly onto each tower leg, while those at the side towers and anchorages are fixed to steel rocker boxes that are free to rotate in the direction of the bridge axis.

The main towers are of steel box construction rising about 156 m above river level and are formed from three fabricated steel boxes that are joined by cover plates to provide a five-cell structure in plan. The legs of each tower are connected by cross members at the top, and just below deck level, and by diagonal stiffened box bracing above and below the deck (Figure 5.3).

The approach viaducts to the bridge are themselves large steel and twin box girders with a reinforced concrete deck slab. The 11-span south viaduct is 438 m long and the 6-span north viaduct is 253 m long.

The design and construction of the bridge is described in detail by Anderson et al. [1]. The maintenance and operation of the bridge and some of the major improvement works carried out are described fully by Andrew and Colford [2].

5.1.3 Role of the Forth Estuary Transport Authority

The Forth Estuary Transport Authority (FETA) is the corporate body responsible for the management, maintenance, and operation of the FRB.

Figure 5.3 Suspended span stiffening truss.

The Authority's purpose is to maintain the FRB in a safe, efficient, and cost-effective manner while minimizing disruption to traffic. FETA was created in 2002, when the former Scottish Executive used powers contained in the Transport (Scotland) Act 2001 to replace the old Forth Road Bridge Joint Board.

The FETA board is made up of 10 elected members from four constituent local authorities as follows:

- City of Edinburgh Council represented by four members
- Fife Council represented by four members
- Perth and Kinross Council represented by one member
- West Lothian Council represented by one member

The chair of the board is rotated between the cities of Edinburgh and Fife every 2 years. The chief engineer and bridge master is the senior full-time official and reports directly to the FETA board. There are approximately 70 personnel employed in bridge maintenance, traffic operations, and administration.

5.2 MAJOR IMPROVEMENT WORKS

5.2.1 Summary of major improvement works

In the first 35 years from opening, several major capital projects have been carried out to replace, strengthen, or improve elements of the structure. These projects were necessary due to changes in traffic loading, design code changes, and deterioration of components and to address risk assessment to evaluate shipping vessels' impact on the bridge.

These projects are listed as follows:

Strengthening of viaduct box girders	£1,145,485
Main tower wind bracing strengthening	£2,916,312
Main tower strengthening	£12,742,999
Construction of pier defenses	£9,914,357
Hangar replacement	£9,706,998

Between the years 2000/2001 and 2013/2014, a total of over £104 million has been spent on bridge strengthening and improvement. The schemes carried out that had a value more than £1 million include the following:

Main-span resurfacing overlay	£1,068,893
Reconstruct toll plaza	£1,323,532
Main-span surfacing south	£3,614,494
New gantry and runway	£3,134,495
Main tower painting	£7,497,980
A8000 approach road upgrade	£16,633,843
Main cable acoustic monitoring	£1,182,215
Main cable first internal inspection	£5,472,439
Toll equipment replacement	£8,331,885
Viaduct bearing replacement	£18,650,176
Replacement of main expansion joints	£3,060,480
Investigation of anchorages	£5,198,514
Main cable dehumidification	£11,532,067
Main cable replacement/augmentation study	£1,023,873
Cable band bolt replacement	£4,374,281

In addition to the improvement or strengthening projects already carried out, painting of the bridge is a high-cost recurring work item. There are almost 270,000 m² of steel surface area on the bridge excluding the main cables and parapets, and maintenance painting is almost a continuous process. Repainting of the suspended span truss, which has over 202,000 m² surface area, is estimated to cost £65 million as the existing chlorinated rubber paints cannot be coated over with modern epoxy-based systems. Therefore, all the existing coatings are likely to have to be removed before a 20-/30-year epoxy system can be applied. This work will also involve extensive access and full containment systems.

5.2.2 First internal inspection of main cables

The main cables of a suspension bridge are the primary load-carrying members; the cables on Forth have been regularly inspected externally and no leaching of water or moisture staining had ever been recorded. However, it was acknowledged that the condition of the individual wires within the cables could not be determined with any certainty.

The United States has many more suspension bridges older than Forth and concerns over the condition over deterioration of strength of the cables due to corrosion led to the Transportation Research Board of the National Academies developing a guide for main cable inspection and strength, published as National Cooperative Highway Research Program (NCHRP) Report 534 [3].

Although the guide was not published until 2005, in 2003/2004 the FETA board decided to adopt the draft NCHRP Report 534 recommendation that aerially spun cables on suspension bridges over 30 years old should be opened for first internal inspection.

This first internal inspection work was completed in 2004 and 2005 and a total of ten 18 m long panels (a panel is the length of main cable between vertical hangers) at various points on both cables were opened, inspected, and rewrapped. In general eight samples of individual wires, each 6 m long, were taken at each panel and cut to a length of 254 mm for tensile testing. Tests to determine the degree of deterioration of the zinc coating and tests on water samples were also undertaken. AECOM assisted by Weidlinger Associates supervised the works, which were undertaken by the contractor C Spencer Ltd.

To inspect a panel, at first, the wrapping wire had to be removed, followed by the protective layer of red lead paste, which had become very friable. Strict containment had to be in place for this work. Following the American practice, initially eight wedge lines were opened around the cable by using brass chisels and then driving in hardwood and plastic wedges. This allowed inspection to take place down to the middle of the cable.

Much to the surprise of the inspection team, fairly extensive corrosion (Figure 5.4) and wire breaks (Figure 5.5), although these were relatively small in number, were found in some panels. Given the relatively young age of the bridge, these results gave cause for concern.

Figure 5.4 Internal inspection methods.

Figure 5.5 Main cable broken wires.

The condition of the wires exposed for the worst panel (midspan on the east cable) is shown in the form of a tree ring diagram in Figure 5.6. The corrosion stage of the wires was determined from the NCHRP Report 534 guidelines, which are shown in Figure 5.7.

The corrosion stage of the wires and the number of broken wires in all the 10 panels inspected are shown in Table 5.1, and as can be seen, there appears to be little correlation between number of broken wires in a panel and the level of corrosion found in that panel.

Based upon the results of this limited intrusive inspection of the main cables, the Authority's consultants concluded that the loss of strength of the cables based on the worst section uncovered was 8%. This estimate of the current cable strength was determined from statistics obtained from the wire testing using the brittle wire model method of analysis as defined in the NCHRP report [3]. The consultants were asked to project the data to determine the likely cable strength over the next 5, 10, and 15 years if no steps were taken to halt the deterioration. It was recognized that the data from such a projection would have to be treated with caution. However, it was predicted that if deterioration could not be halted, the cable could lose 13% of original strength by 2014 and 17% by 2019.

The main cables of suspension bridges have traditionally been designed using the working stress approach. In the original design, the direct stress in the main cables on Forth was limited to 40 tons/in^2 (618 N/mm^2) and a minimum wire tensile strength of 100 tons/in^2 (1544 N/mm^2) was specified. This would result in a factor of safety (FOS) of $100/40 = 2.50$ and it should be noted that this factor is against ultimate failure. It is usual when designing other structural steel elements using working stress methods to use yield or 0.2% proof stress (which would be 75 tons/in^2 or 1158 N mm^2) rather than ultimate strength to determine the FOS. In the case of the cables, if the 0.2% proof stress were used, the FOS would be $75/40 = 1.875$.

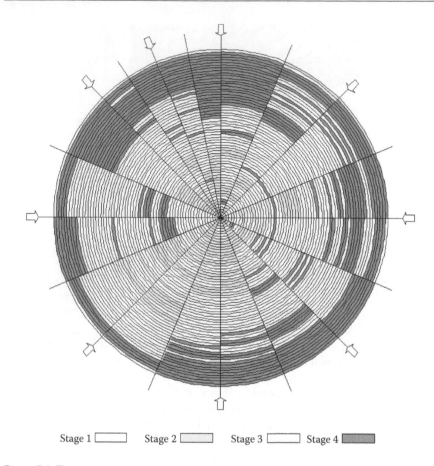

Stage 1 ☐   Stage 2 ☐   Stage 3 ☐   Stage 4 ■

Figure 5.6 Tree ring cross section.

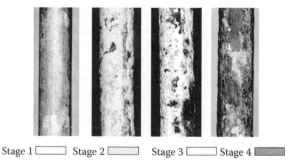

Stage 1 ☐   Stage 2 ☐   Stage 3 ☐   Stage 4 ■

Figure 5.7 Wire conditions as per NCHRP guidelines. (From Mayrbaurl, R., and S. Camo, NCHRP Report 534 Guidelines for Inspection and Strength Evaluation of Suspension Bridge Parallel Wire Cables, Transportation Research Board, Washington, DC, United States, 2004. With permission.)

Table 5.1 Conditions of wires in cable panels that were inspected

| Cable panel | Location on cable | Corrosion level | | | | Broken |
		Stage 1	Stage 2	Stage 3	Stage 4	
1. E100S-100N	Low	0%	8%	55%	37%	8
2. E00N-02N	Low	9%	86%	4%	1%	0
3. W00S-02S	Low	11%	73%	10%	6%	31
4. W100S-100N	Low	0%	36%	38%	26%	17
5. W00N-02N	Low	11%	75%	6%	8%	8
6. E100N-98N	Low	0%	16%	58%	26%	8
7. W22N-24N	High	0%	45%	42%	13%	5
8. E18S-20S	High	19%	60%	19%	2%	0
9. W76S-74S	High	7%	58%	25%	10%	2
10. E60N-58N	High	22%	43%	26%	9%	7

In order to determine the 2004 value of the FOS, W.A. Fairhurst & Partners was asked to conduct an assessment of the actual current dead and super dead load of the bridge, and verify the AECOM work. Super dead load (or superimposed dead load) is the weight of materials and elements that are not structural elements (e.g., road surfacing, rail ballast, parapets, lamp standards, and ducts). It was concluded that the bridge was about 3.5% lighter than the engineers had estimated for the original design and that the actual original FOS against ultimate failure was nearer to 2.59.

The live load used in the analysis was the 2002 Bridge Specific Assessment Live Load, and the bridge-specific footway loading assumed was a low value of 0.15 kN/m. Assuming a strength loss of 8%, the 2004 FOS was estimated to be 2.27 and was projected to fall below 2 between 2014 and 2019. This is illustrated in Figure 5.8. It should be noted that in order to determine the FOS, the panel with the largest loss of strength was used but the load adopted was that found in the highest loaded panel. This was a conservative approach but was justified given that only 10 out of 198 panels were inspected.

There is no absolute number that determines the minimum acceptable FOS of the main cables of a suspension bridge. NCHRP Report 534 refers to a FOS of 2.15, and based on discussions with some U.S. bridge authorities, the feeling was that a number of owners would be uncomfortable running uncontrolled traffic with a value less than 2 as a permanent condition.

The results of the first internal inspection of the main cables at Forth were not only significant for the bridge authority but also to the wider bridge community. The results suggested that there were serious doubts over the use of paint systems to try to protect the cables of suspension bridges. As a consequence of the work at Forth, the owners of the Severn and Humber Bridges in England instigated cable inspection programs.

At Forth, it was concluded that if the rate of deterioration due to corrosion could not be halted and consideration would have to be given to the possibility of introducing loading restrictions on the bridge around 2014.

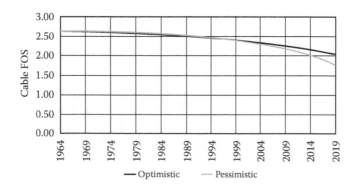

Figure 5.8 Estimate of future FOS of the main cables after the first inspection.

Given the significance of the crossing, both strategically and locally, this news caused quite a political and media stir within Scotland.

Following the findings, in November 2005, the Scottish executive appointed Flint & Neill Partnership assisted by New York–based Ammann & Whitney to audit the results. The purpose of the audit was to carry out a desk study of the findings and to advise whether they were reached using a process of appropriate rigor and whether the conclusions were reasonable.

In January 2006, a comprehensive audit report was submitted and concluded that the initial internal inspection and cable strength calculation had been carried out in accordance with accepted practice in the United States and in general conformance with NCHRP guidelines. It was noted in the audit that certain assumptions had been made that differed from the NCHRP methods but they were reasonable and appropriate considering the particular characteristics of the FRB versus the older U.S. bridges upon which the NCHRP study relied on as case studies.

The total cost of the project to inspect a relatively small length of the cable was £2.3 million. A full description of this work is given by Colford and Cocksedge [4].

It was clear that action was required to halt or limit the corrosion and to monitor the cables. The following works and studies were commissioned for this purpose:

- Installation of acoustic monitoring on both cables
- Installation of a dehumidification system on both cables
- A feasibility study to determine whether or not the cables could be replaced or augmented if the corrosion continued and to investigate the condition of the main cable anchorages

The audit report from Flint & Neill supported the adoption of these proposals.

5.2.3 Acoustic monitoring

In November 2005, a contract was awarded to a French contractor, Advitam SAS, to fit both cables with acoustic monitoring equipment supplied by Pure Technology (Figure 5.9). The system was designed to provide continuous monitoring of wire breaks in the cables in order to increase confidence that the worst section of cable had been uncovered. Site works commenced in April 2006 and the system was operational in August 2006. The total installation cost was £620,000, with monitoring costs of £55,000 per annum. This work was supervised by AECOM. Since commissioning, to date (December 2013) a total of 83 events have been determined as wire breaks and 7 events as possible wire breaks. However, given that there are 11,618 individual wires in each cable, it is not considered that this activity is significant enough to warrant action except to continue to monitor the situation and review it

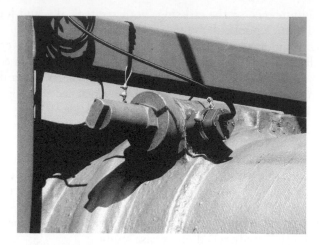

Figure 5.9 Acoustic monitoring.

over time. Information on wire breaks is useful intelligence when selecting panels to open during the next internal inspection of the cables.

The system installed in 2006 has some limitations. The system on FRB has only 15 sensors on each cable and there is a concern that wire breaks may be being missed. In addition, it was designed to be a wireless system that had to be installed as hardwired as the regulatory authority in the United Kingdom would not permit use of a suitable wireless band for transmission of the signal.

Given this, the bridge authority is proposing to replace or upgrade the system in 2014. The new system is estimated to cost £750,000 and will include between 56 and 59 sensors on each cable, approximately one every second hanger. The aim of the project is to increase the likelihood of detecting wire breaks in the main cables. Given the history and work carried out on the main cables and the estimated reduction in strength that has been established, it is important to maintain confidence in wire break detection.

5.2.4 Dehumidification

After discussions with other bridge operators in Europe, Japan, and the United States, it was decided to install a system of dehumidification on FRB. This work involves pumping dry air into the cables at various points after wrapping it first in an airtight neoprene membrane.

Dehumidification is a well-tried system for preventing corrosion of steelwork. Although it was already being used in box girders of some bridges and indeed in the anchor chambers at Forth, its application to main cables of suspension bridges was relatively new. Such systems have been fitted to younger bridges in Japan, Sweden, and Denmark where corrosion has been uncovered.

The only viable option for bridge owners to stop cable corrosion prior to the development of dehumidification was to oil the cables. While a small number of main cables on bridges in the United States have been oiled, the results have not been uniformly successful. The cost of unwrapping cables, oiling, and rewrapping is significant and the potential disruption to traffic also has to be considered.

As already described above, the main cables are made up of 11,618 parallel wires that have been compacted. However, within the cross section of the cable between the contact points of the wires, there are voids and at Forth these voids make up 20.5% of the cross-sectional area. The key objective in dehumidification is to fill the voids with air with low relative humidity. It is considered that a piece of galvanized wire in a chamber with a relative humidity less than 40% will not corrode. Therefore, if air with a relative humidity under 40% can be introduced into the cable and surround all the wires, the capacity for further corrosion is unlikely. The project to complete the dehumidification of both cables was successfully completed before the target end date of October 31, 2009. Dehumidification of the west cable commenced in March 2008 and has produced the expected slow and steady fall in the relative humidity within the cable.

Figure 5.10 shows the air that is introduced within the cables at Forth is at a very low pressure (about 3000 Pa) via inlets spaced along each cable. The air vents after traveling along the cable either 160 m in the main span or 200 m in the side spans.

On Forth, the air is not taken directly from the atmosphere but is introduced via three plenum chambers, one on each span, located between the west footpaths and the northbound carriageway. The chambers contain the dehumidifiers and the pumps; Figure 5.11 shows a typical chamber.

Figure 5.10 An injection sleeve from cable crawler.

Figure 5.11 Inside a plenum chamber.

Initially, the air was introduced at a relative humidity of 20% to try to speed up the rate of drying. When air with a relative humidity below 40% was vented at the exhausts, the system was run at a relative humidity of 40%. By March 2013, relative humidity levels of less than 40% were recorded over 90% of the length of both cables. However, in some areas near the tower tops, it proved more difficult to reduce the relative humidity to below 40%.

The work was designed and supervised by AECOM assisted by Nippon Steel and was carried out on site by C Spencer Ltd. The final project cost of the dehumidification including consulting engineers' fees was over £11.5

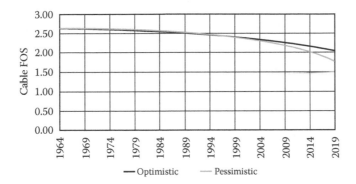

Figure 5.12 Estimate of future FOS of the main cables after the second inspection.

million. The maintenance cost of the system at FRB is around £25,000 per annum and the operating costs, mainly power, are around £30,000 per annum.

It should be stressed while it is considered that there is good reason to have confidence that dehumidification could slow down or halt corrosion, there is no body of evidence yet to assure that this will work on Forth. While it is fairly certain that dehumidification will stop further corrosion, it remains to be seen what will happen to already cracked and damaged wires.

The reasoning for this is that individual wires are likely to contain some microcracks along their length. The number and depth of these cracks and their potential to grow within the dehumidified cable are unknown. Therefore, the determination of how many of them will, in turn, lead to wire breaks cannot be determined. As loss of strength is directly related to wire breaks, the future strength of the cables is unknown. The data from the dehumidification and acoustic monitoring systems are kept under review. These data, along with future intrusive cable inspections, will help provide evidence of the effectiveness of the dehumidification system.

Prior to the system being commissioned on the east cable, a full internal inspection of three panels at midspan was carried out in 2008 to benchmark the condition of the cable prior to starting dehumidification and to determine the strength loss in the cable since 2004/2005. This limited inspection showed that the strength of the cables had dropped further and the loss of strength was around 10% just before dehumidification was installed. This is shown in Figure 5.12.

5.2.5 Main cable strengthening feasibility study

There is no absolute guarantee that dehumidification would prevent further deterioration and, therefore, further strength loss. Hence, a feasibility study was commissioned to determine whether or not the main cables on

Forth could be either replaced or augmented. In August 2006, following a quality/price tendering exercise, W.A. Fairhurst & Partners were awarded the work to carry out the feasibility study. The team included engineering consultants COWI Consult (Denmark), Ammann & Whitney (United States), traffic modeling consultant SIAS (Edinburgh), and economic consultant Roger Tym & Partners (Glasgow).

It had been recognized that should it become necessary in the future to consider replacing or augmenting the main cables, several significant engineering difficulties had to be overcome. These included the following:

- The main towers and existing cable saddles have already been strengthened and new load paths have to be identified at the tower tops.
- The existing cable anchorages may need to be replaced or augmented.

In addition, an assessment of the stiffening truss had concluded that several of the members were overstressed.

The extent of the carriageway restrictions that would be required to carry out replacement or augmentation of the main cables and the effects of those restrictions on the adjacent road network were also determined by the study.

It is known that main cables had been supplemented on Tagus Bridge in Portugal and replaced on Tancarville Bridge in France while both bridges retained some operational capacity. It was also accepted that replacement or augmentation of the main cables of a suspension bridge carrying live traffic involves some element of risk to users. Although these risks can be managed, they cannot be eliminated. The only way to eliminate risk to users would be to close the bridge completely to all traffic over the work period, which was estimated to be between 3 and 4 years. Without an alternative means for traffic to cross the Forth at Queensferry, total closure of the bridge for the duration of the works was not considered further. Therefore, the study had to balance the need to keep the bridge operational for users, with the assessed risk to those users from carrying out major construction works directly above them.

The risk assessment was key part of that study and developing a risk register was the first part of that assessment. An important first step of any risk assessment is determination of the client's risk appetite. Attitudes toward risk can vary from country to country, which proved interesting with a multinational team working on the project.

Three main structural options were identified following a selection process and these are given as follows:

- Option A—replacement above: New cables with sufficient capacity to carry all the loads would be constructed above the existing main cables. Once all the loads are transferred to the replacement cables, the existing main cables would be removed from the bridge. The duration of the contract is 8 years.

- Option B—augmentation above: This is similar in many respects to option A except that the existing main cables would be retained and the load from the suspended structure shared between the cables. The duration of the contract is 7 years.
- Option C—augmentation to side: This option consists of new cables constructed to the side of the existing main cables, with the load from the suspended structure shared between the new and existing cables. The duration of the contract is 7 years.

The augmentation-to-the-side scheme (option C) would involve the provision of a cable smaller than the existing. However, in the augmentation-above scheme (option B), a cable of size similar to that of the existing cable could be supported. The difficulty in any augmentation option was the sizing of the cable. There would have to be a high level of confidence in the prediction of the long-term strength of the main cables before any augmentation scheme was adopted. This high level of confidence could be achieved only after future inspections and subsequent determinations of cable strength.

The estimated design and construction costs of each option are shown in Table 5.2. It should be noted that these are 2007 costs, and while they incorporate elements of contingency, they do not account for inflation or optimism bias.

These options require different traffic management layouts that would involve lane and carriageway restrictions on the bridge over a significant number of weeks and total bridge closure on a number of weekends and overnight.

It was recognized that these closures on the bridge would have an immediate effect on traffic and cause significant disruption. Traffic modeling was developed as part of the study to examine the economic effects of the restrictions required for each of the options. This modeling was used in conjunction with the traffic model for Scotland, developed by Transport Scotland. The notional costs of delaying vehicles in roadworks were derived from the Department for Transport's Transport Analysis Guide and were combined with the results from the modeling to produce an estimate of the notional total cost of the delays. If a carriageway is closed on the bridge on a weekday, the resulting costs of notional travel time delay were estimated

Table 5.2 Design and construction costs

Main structural option	Design and construction costs
A—replacement above	£122 million
B—augmentation above	£120 million
C—augmentation to side	£91 million

to be on the order of £650,000 per day. It was noted, though, that this was a worst-case scenario as no mitigation measures had been considered. The estimated costs of travel time lost are shown in Table 5.3 for each option.

The study team also looked at using off-peak carriageway closures, as these have the least disruptive effects on the network. The use of off-peak carriageway closures does increase the total number of weeks of closures required but the delay costs are reduced. However, there are issues with risk management and the actual daily physical alteration of the traffic management to suit the traffic flows.

In addition to these costs due to travel time delays, it was also recognized that there was a wider economic cost to business caused by major maintenance work on the bridge, and an assessment of these costs was also part of the study. A business survey was carried out in November/December 2007 to obtain feedback from the business community in Fife, Edinburgh, Lothians, and the Forth Valley on the effects of major bridge maintenance works and associated traffic management restrictions on business and employment.

Determining these wider costs was not an easy task. Brevity and clarity are the key principles in conducting any business survey, and it was decided that presenting various structural options with their associated traffic impacts would lead to overcomplication and, consequently, a dilution of response. Therefore, the survey sought to gauge the impact on business performance of major maintenance on the bridge, resulting in lane and carriageway closures, over a sustained period.

The study conclusions were that major maintenance work on FRB over a sustained period, involving lane and carriageway closures, would potentially result in the following:

- Economic output falling to a level on the order of £1 billion below that anticipated were the bridge to be operating normally
- A drop in turnover in excess of £1.3 billion below that anticipated were the bridge to be operating normally
- A loss of around 3200 jobs, some of which may turn out to be permanent

These figures were obviously very substantial and the effects on business would be most strongly felt in Fife, including the negative effect on employment. There would also be an adverse impact on the environment

Table 5.3 Estimated costs of travel time lost (base year: 2007)

Main structural option	Costs of travel time lost
A—replacement above	£335 million
B—augmentation above	£238 million
C—augmentation to side	£212 million

for each of the options with increased vehicle emissions associated with the increased delays.

The study, which was concluded in early 2008, showed that while it was feasible to replace or augment the cables on FRB and retain some operational capacity, there was a residual risk to users and the cost and disruption to traffic and business would be significant.

5.2.6 Main cable—Third and post dehumidification internal inspection

The cables were inspected and assessed following the NCHRP Report 534 guidelines [3] during April to September 2012. This third internal inspection of the main cables was carried out in order to

- Verify that the dehumidification system was protecting the cables and determine the estimated current and future strength of the cables;
- Look at other high-level panels not previously inspected where the cable forces are highest;
- Investigate locations that took longest to dry out and that had exhibited the most acoustic emission activity; and
- Revisit a previously inspected panel to investigate any further deterioration since 2008.

Following a tendering exercise, the consulting engineers, Flint & Neill, in conjunction with Ammann & Whitney and COWI, were awarded the project on behalf of the authority in November 2010. AECOM was also selected to review the inspection work.

In order to reduce costs, following a government spending review, the adjustments listed in the following were made to the main contract:

- The number of panels inspected was reduced from 12 to 8.
- The authority took more of the risk associated with the weather.
- Modifications to the access platforms were rationalized.
- The off-site testing work was procured out with the main contract.
- The authority purchased the replacement wire and wire products directly.
- The site compound was situated in Rosyth rather than Builyeon Road.
- The extent of the independent review was reduced.

In October 2011, the authority approved acceptance of a negotiated tender that reflected these adjustments submitted by C Spencer Ltd., for the sum of £2,573,310, to carry out the third internal inspection of the main cables. The contract was completed on site in October 2012 and was on time and under budget. The final contract cost was £2,560,824.

After inspection, wire sampling, and laboratory testing, an analysis of the condition of the cables was carried out by Flint & Neil and a review of that assessment was carried out by AECOM. The findings from the inspection were as follows:

- The total number of panels inspected in the years 2004 to 2012 was 18 but this is a smaller number than that recommended by the NCHRP guidelines. This was mainly due to budget constraints.
- Broken wires were found in six out of eight panels inspected during the third inspection. The largest number of broken wires found was 16 and no correlation could be made between broken wires and the number of stage 4 wires found.
- The highest percentage of stage 4 wires was found in the panel near the mid–main span (12%). Significantly high numbers of stage 3 wires were found in six out of eight panels; a particularly high percentage of stage 3 wires (71%) was found near mid–main span. A total of 55 broken wires were found.
- The inspection revealed that in total 45% of the test specimens failed below the specified minimum tensile strength of 1544 N/mm^2.
- Fatigue tests at an elevated stress range revealed that only 27% of the total number of specimens tested passed two million cycles. It is reasonable to conclude that the fatigue strength significantly decreases as the steel corrosion advances, and the majority of failures were due to a corrosion pit and presence of a preexisting crack.
- Zinc-coating tests indicated that there was an adequate thickness of galvanizing on all the stage 1 specimens and that they all exceeded a zinc weight of 275 g/m^2.
- Some of the cable panels have been inspected by two independent teams, Flint & Neill, the assessing engineer, and AECOM, the reviewing engineer. Both teams have been involved in the inspections of all the major suspension bridges in the United Kingdom and have international experience.
- A cable panel that had been found to contain high levels of corrosion and the most number of broken wires was reinspected. As there is a degree of interpretation in the inspection process, this panel was inspected by the two teams. No broken wires were found on this occasion and the Flint & Neill engineers found fewer stage 4 wires. The AECOM team found a similar number to those found previously in 2008. Overall, it appears that the level of corrosion in the reinspected panel has not increased.
- The inspection gathered more wire strength data, which have been used to expand the knowledge of the cable capacity. The variation that these data produce was comfortably within the band of uncertainty expressed in the FOS used for determination of the safe working limits for the bridge.

- The result of this finding was that the likelihood of the areas of greatest corrosion being found at the lowest points of the cable had been affirmed and that future inspections should be targeted in these regions. It is much easier and cheaper to inspect in these areas. Hence, the finding is significant in terms of reducing the inspection costs while not compromising structural safety. The higher points, where the loads are greater, were inspected in greater number in the 2012 inspections and found to have lower levels of corrosion.
- The number of broken wires found in each of the three inspections was considered directly comparable as different panels had been chosen each time. Early inspections targeted the likely worst panels with most wire breaks and later inspections looked at other areas on the bridge that had the highest load and where existing data were limited.
- Further data have been gathered from new panels from different areas on the cables and new calculations of the remaining strength of the cables using different methods were made.
- Within the bounds of the limited number of panels inspected and small differences in the wire inspections, sampling, and testing, Flint & Neill found that the FOS of the cables has not materially diminished and that gave strong comfort that the newly installed dehumidification system was retarding the corrosion of the bridge wires.
- Although wire corrosion appears to have been slowed down, there are still existing cracks in some of the wires that might propagate from corrosion pits that existed prior to installation of the dehumidification system. These cracks may eventually lead to wire fractures but the rate of breakages is expected to slow significantly. Given this, it is expected that the FOS of the main cables will not diminish significantly in the future as long as the dehumidification system continues to function effectively.
- Therefore, on the basis of the NCHRP guidelines and the findings from this latest inspection, the next intrusive inspection of the FRB cables should follow in 5 years time in 2017. However, the guidelines do not take account of the dehumidification system nor the knowledge gained from ongoing acoustic monitoring. Therefore, it is recommend that a further intrusive inspection of a few lower panels is tentatively planned for 2017 but confirmed nearer the time when the ongoing monitoring has been taken into account.

AECOM reviewed Flint & Neill's findings and made no significant comments.

As noted above, although wire corrosion appears to have been slowed down, there are still existing cracks in some of the wires that may eventually lead to wire fractures. This should result in a further slowing down of deterioration of the cables and lead to a reduction in the loss of magnitude in the FOS. It will be important for the future resilience of the cables that the dehumidification system is maintained and the existing acoustic monitoring system is augmented.

Flint & Neill has recommended that a further intrusive inspection of a few lower panels be tentatively planned for 2017 but will be confirmed in the future based on the data from the ongoing monitoring.

A degree of uncertainty concerning the magnitude of future strength loss of the main cables will always remain. Thus, cables will require to be continually monitored and be subject to a regime of internal inspections and strength evaluations for the remainder of the service life of the bridge. However, the results of this latest inspection, albeit reduced in scope, were encouraging. The dehumidification system applied to both cables appeared to be slowing down the rate of deterioration.

5.2.7 Further research on main cables

In May 2012, the U.S. Department of Transportation's Federal Highway Administration published *Primer for the Inspection and Strength Evaluation of Suspension Bridge Cables* [5]. The main purpose of the primer was to supplement NCHRP Report 534 [4]. However, it also presented an alternative methodology to that presented in NCHRP Report 534 [3]. This alternative is the BTC method [6], by Bridge Technology Consulting (BTC), a probability-based method of evaluating the remaining strength and service life of main cables. The authority is currently evaluating a limited BTC method study against the outputs from the NCHRP work on Forth in a limited desk study.

5.2.8 Investigation of anchorages

The anchorages of the main cables of a suspension bridge are critical elements of the structure. At FRB, tunnels were formed within the rock at each of the four anchor points and filled with concrete. See Figure 5.13.

The main cable wires splay out in the anchorage chambers and loop round strand shoes, which are, in turn, bolted to the face of the concrete tunnels; see Figure 5.14. Friction between the concrete tunnel and the rock, and the weight of the overburden above, prevents the cables pulling the concrete tunnel out of the ground. The concrete in the tunnel itself is not strong enough to withstand the forces from the cables and was strengthened using posttensioned galvanized high–tensile strength steel wire tendons. The tendons consisted of four 32 mm diameter strands in a 100 mm diameter steel duct, which was then filled with grout. The strands are made up mainly of 6 mm diameter galvanized wires. There were 114 tendons per anchorage and each tendon was pretensioned with a 152-ton load. This use of pretensioning in the buried concrete anchorage tunnels at Forth was considered innovative at the time. Unfortunately, this form of construction

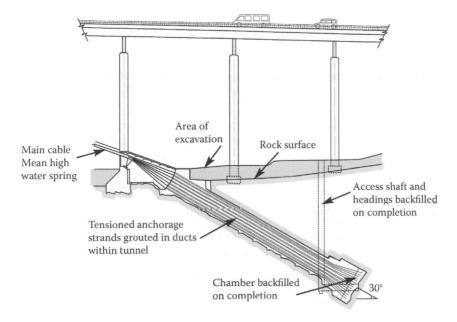

Figure 5.13 Section showing an anchorage.

Figure 5.14 Anchorage strand shoes.

can be vulnerable to corrosion and deterioration, especially in a saline environment such as is found at Forth.

Records and papers acquired around 2005 relating to the construction of the existing anchorages highlighted various problems during construction, particularly in relation to early depletion of the galvanization protecting the posttensioning strands that are housed in grouted ducts set in the concrete tunnel.

The reports on the issues and conditions encountered during construction determined the need to carry out a special inspection or investigation to establish the condition of the posttensioning strands. Frequent inspections are carried out to monitor for movement within the anchorage chambers at the tunnel/strand shoe interface and no signs of distress or movement had ever been recorded. Nevertheless, it was considered that the investigation was necessary to ensure the long-term structural integrity of the anchorages and was also considered to be a proactive measure to ensure that all accessible parts of the structure were inspected.

There is guidance from the United Kingdom's Department of Transport for inspecting posttensioning in bridges as it is acknowledged that there can be problems with this type of construction. The guidelines refer mainly to the difficulties in establishing the condition of posttensioning strand in bridge decks. These difficulties are exacerbated in a tunnel. After much discussion, research, and consultation with various specialists, it was concluded that three separate methods of investigation should be taken forward. These were as follows:

- Excavation behind the anchorage chambers down to the top of the tunnel to expose and inspect the posttensioning strands.
- Full-scale testing of a number of the sockets within the anchorage chamber.
- Other nondestructive testing methods, such as acoustic monitoring, have been examined but these methods have not appeared to show much potential in solving the problem alone but may provide some useful data when combined with the first two methods.

Fairhurst was engaged to carry out the work on behalf of the authority and Flint & Neill was appointed to chair a peer review panel in order to audit and review the work by Fairhurst. Flint & Neill was assisted by the New York consultants Ammann & Whitney and the chief bridge engineer of Trunk Roads Network Management of Transport Scotland.

Initially, it is proposed to commence work at the south anchorage as it was expected that if corrosion of the strands had occurred, it would be worse at this location. The excavation work and visual examination and testing of the strands would be carried out in the first instance as a separate contract. Further work would depend on the results of this examination.

The excavation was expected to be difficult as the ground conditions vary and there is methane present within the shale. The work was further complicated by the close proximity of the foundations to the viaduct piers. Any in situ testing of the anchorage sockets would also be challenging, as access is very difficult within the anchorage chambers. Safety of the workforce, the bridge, and users governed all aspects of this complex project.

Work on the anchorages was not anticipated to involve major disruption to bridge traffic. However, given the nature of the work, significant environmental issues were expected, especially in regard to noise, dust, and discharge from the excavation and hydrodemolition. To assist the community, it was determined that a requirement of the contract was to ensure that the contractor named a member of staff as community liaison officer. Prior to the commencement of the works, a number of meetings were held with local residents and the community councils to provide as much information on the works as possible and to establish points of contact for the duration of the works.

A sum of approximately £7.3 million had been set aside for the project. However, this was intended to cover excavation and some limited in situ testing work on the south anchorage only. In April 2011, the authority approved the acceptance of a tender submitted by John Graham (Dromore) Ltd. to carry out the works to investigate the condition of the tensioned strands within the southern anchorage tunnels. The investigation of the anchorages commenced on site in August 2011 and was scheduled to be completed before the winter of 2013.

The contract with the contractor was a standard-form New Engineering Contract (NEC) 3 option C target cost contract with a tendered target cost of £3,497,849. There was a pain/gain element to the works. If the final cost of the works was above the target cost, then the contractor pays a share of that increased cost. Conversely, if the final cost is below the target cost, then the contractor receives a share of the saving. It should be noted that the target cost could be adjusted during the contract to account for unforeseen work.

The contract involved excavating down through overburden and rock to the top of both tunnels to expose the concrete forming the tunnels over a length of around 10 m. This work was carried out in a carefully controlled manner, utilizing a combination of mechanical excavation and hydrodemolition techniques. Side slopes were strengthened by subsurface soil nailing and rock anchors. Further, careful hydrodemolition was used to remove the tunnel concrete to expose a number of steel ducts housing the pretensioning steel strands. These ducts were then carefully cut open and the grout washed out to expose the strands. Once exposed, the strands were inspected and tested in order to allow an evaluation of the capacity of the anchorages. It was initially proposed to build permanent access chambers to aid future monitoring prior to the site being reinstated back to the original ground level profile.

If the investigation revealed significant deterioration in the steel strands within the anchorage tunnels, then, depending on the level of that deterioration, the authority was aware that measures may have to be considered to limit loading on the bridge. Full-scale testing of the sockets within the southern anchorage chambers was not carried out.

The work progressed well on site, and by September 2012, a number of ducts were visible at the crown of the southwest anchorage tunnel and their external condition was excellent. Radiographic techniques were used, prior to cutting the ducts open, to determine the position of the strands within the ducts.

However, from examination of the ducts first exposed, it was apparent that they were some 400 mm deeper and at a steeper angle than recorded on the as-built drawings. Added to this, the rock head was higher than expected and the extent of ancillary foundations from the original construction, which had to be removed, was also greater than anticipated. All of this meant that the contractor had to spend far longer than originally programmed and additional resources required led to additional cost.

The original proposal was that the anchorage tunnel would be excavated to a level in order to maintain a minimum of 500 mm cover over the post-tensioning ducts. As the condition of the tendons/strands was unknown, and it was considered that there could be significant deterioration within the strands, the construction of permanent inspection chambers was included in the contract to allow for future monitoring. Additionally, the chambers would also have provided a more controlled environment for the current investigation work.

When reviewing the work to remove the existing tunnel concrete, the contractor proposed that, on both safety and economic grounds, to remove all concrete from around the ducts prior to the construction of the chambers. Temporary protection was proposed to cover the ducts prior to the chambers being constructed. This early exposure of the ducts offered the opportunity to view the strands a lot earlier than that had been programmed, albeit over a more limited length, and in November 2012 the first ducts were carefully opened in the east tunnel.

The external surfaces of the ducts were in very good condition. The ducts are not galvanized and are reliant on the alkaline nature of the concrete to protect the steel. The welded joints were fabricated to a good standard and bitumen strips around the welds were still evident. This finding was taken to be indicating that the strands were well protected by the anchorage concrete and that the anchorage was not allowing the ingress of water. The next stage was to carefully cut open a length of the duct to expose the grout, wash it out using high-pressure water, and expose the strands for inspection. To try to ensure that strand wires were not cut when opening the ducts, radiographic techniques were used which proved successful.

Where exposed, the strands were found to be in remarkably good condition with no visible signs of corrosion of the wires. Where access permitted, the diameter of the strands was checked and no evidence of a change in diameter was found, suggesting no internal corrosion had occurred. Careful wedging using hand tools and hardwood wedges was used to separate the strands to enable a visual inspection to be made around the circumference of the strands; see Figure 5.15. The dull gray color of the galvanizing on the surface of the strands indicated that oxidation of the zinc coating had occurred and localized darker areas on the strands were noted, suggesting

Figure 5.15 Exposed pretensioned anchorage strand.

that these areas had oxidized to a greater degree. However, overall the strands that were inspected were in good condition, with no evidence to suggest deterioration from the time they had been grouted.

On one of the ducts, the strands were found to be misaligned and on one strand the lay of the outer wires was out of alignment, with one wire bulging up clear of the rest of the strand. The strands all appear to be loaded and were realigned within approximately 1 m of the top circumferential cut. Inspection revealed that the displaced wire and the strands were fully grouted, with grout visible between wires and strands. This indicated that the damage occurred during installation of the strand and prior to the grouting operation.

A temporary acoustic monitoring system was installed within the anchorage chambers on some of the strand sockets and anchor plates to monitor for activity during the investigation. Only one event was recorded during the 18 months of the investigation and that was not associated with ducts being opened. Whether or not that event represented a wire break could not be confirmed.

Nine ducts out of a total of 114 were opened in each tunnel of the southern anchorage. Those were around 8% of the posttensioned ducts opened in this investigation. In comparison, only 4% of the main cable panels were opened during the 2012 main cable inspection, and from the visual inspection, the condition of the wires in the anchorage strands is significantly better than the condition of the wires inspected in the worst panels in the main cables.

By the end of January 2013, all the strands within the 18 ducts were exposed and all were found to be in a similarly good condition. Given the good condition of all the strands exposed, a hold point was established to determine whether or not further investigation work was going to provide significantly different information, balanced against the further removal of ducts and grout, which appears to be providing very good protection to the strands. The further work that was considered by Fairhurst, the peer review team, and the authority was as follows:

- Sampling of strands/wires
- Further radiography
- Magnetostriction
- Load testing

There was agreement between Fairhurst, Flint & Neill, and FETA that, given the unexpectedly good condition of the strands, sampling of strands/ wires, further radiography, and magnetostriction testing were no longer required. Fairhurst considered that it would be worthwhile undertaking load test to establish the measure of the load in the strand, although these tests would be limited in extent. However, the peer review team's comments on the proposed load testing were as follows:

- The proposed test would be very difficult to carry out and probably inaccurate without calibration of the equipment in a laboratory with a strand of the same age and construction and would risk breaking the bond with the grout over an unknown length.
- The tests would provide little benefit to the assessment of the overall strength of the anchorage and would simply affirm the initial pre-stressing force in the strands.
- As there is no evidence of movement in any of the anchorage plates within the anchorages, there is no reason to suspect that any plate was not properly stressed.
- Load testing would only show that the force in a strand might have reduced; any loss of section would not result in a reduction in load but would give rise to an increase in stress commensurate with the loss of area.

Given the above, the benefits of the output from further investigation, including load testing, were considered not to outweigh the risk of disturbing the bond at the end of the opened section of duct.

The investigation had been limited in nature, and it was not possible to access the wedging area at the bottom of the anchorage tunnels. In addition, it has not been possible to reconcile the apparent concerns with hydrogen emissions noted during the original construction (and the possible consequences for loss of galvanizing on the strands) and the general good condition of the galvanizing and strands themselves noted during the inspection.

The possible source of hydrogen emission was considered further. However, in the absence of any significant loss of galvanizing on the strands, it was postulated that the hydrogen gas and corrosion products observed at the time of construction arose from the corrosion of the zinc metal at the boundary of the socket and zinc socketing material. This interface would not necessarily have been blocked completely after casting of the zinc within the socket as the zinc shrinks slightly when cooling. During construction, the anchorage tubes were open to water prior to grouting, and with any small gaps in the socket head, a path for corrosion was available. However, as has been witnessed from the anchorage excavations, the grout was effective in blocking the strands and there is no evidence of continuing corrosion at the anchors. Any gaps between the zinc and the socket would be plugged by the corrosion products. This is not considered to create vulnerability in the structure as the mass of zinc in the socket is very substantial.

It is recognized that this was just a postulate, but one that is supported by the data and the recent inspections, and could not have been done without the excavation. The only further additional investigations that could have been carried out were the wholesale destructive testing of the anchorage around the anchorage plates. The cost of this would be prohibitive and, given the lack of evidence currently available as to which anchorages were observed to be bleeding off hydrogen gas, this is as likely to imperil a sound anchorage as to discover evidence of corrosion in another.

As a consequence of the above, after careful consideration, it was determined that enough evidence on the condition of the anchorage strands has been gathered during this investigation to conclude that anchorages on the southern bridgehead were in a satisfactory condition. Based on the information available from the investigation, it was considered that the risk of the structural integrity of the southern anchorages being compromised has been reduced significantly as a result of these findings.

In addition, given the history of the construction of the anchorages both north and south, and the better conditions on the north side, it was recommended that no investigation work be carried out at the north anchorages. From the information obtained from the investigation, it was also concluded that the risk of having to replace the anchorages during the remaining service life of the bridge was relatively low.

Decisions then had to be made about the reinstatement of the anchorages and two options were selected for consideration. These were as follows:

- Reinstate ducts and grout followed by sequenced mass concrete infill: Reinstatement of the ducts and the grout under pressure, along with reconcreting of the anchorage tunnel, replicates the existing protection, which has been shown to have maintained the strands in a good condition for 50 years. There would be no future access to the strands.
- Construct limited chambers over exposed length of ducts: One of the key considerations when the chambers were first proposed at the design stage was that the strands were likely to be in a relatively poor condition and would require future inspection and monitoring. Good condition of all the strands exposed during the investigation raised

questions over the need to build these chambers. The chambers add significant cost not only during construction but also to the future operation and maintenance.

After some discussion, it was decided that building the inspection/ monitoring chambers would not be taken forward as their purpose was to provide the means to monitor a damaged strand, one that might need intervention at a later date. It was determined that the exposed strands have demonstrated the value of the corrosion protection system in the inspected region. They have survived for almost 50 years and there is no reason to presume that they would not continue to provide protection for a similar length of time, assuming that the corrosion protection system could be reinstated. The proposal to reinstate the grout under pressure was considered to provide the necessary protection to the strand, followed by controlled reconcreting of the anchorage shaft locally.

Reinstatement of the excavation and reconcreting around the tunnel was completed in August 2013. The unexpectedly good condition of the anchorages strands exposed also led to a significant savings on the project, estimated to be £0.892 million.

The authority's own staff continues to inspect and monitor the anchorage sockets and plates within the existing anchorage chambers, on the north and south sides, as part of the ongoing inspection of the bridge. The frequency of those inspections reflects the criticality and vulnerability of the anchorages and takes account of the investigation results. Some further statistical analysis of the results of the investigation was carried out by Fairhurst assisted by staff from the University of Strathclyde, who had set up of a statistical model for the investigation.

5.2.9 Viaduct bearing replacement

On the north and south sides of the suspension bridge, there are multispan approach viaducts. The 11-span south viaduct is 438 m long, while the 6-span north viaduct is 253 m long. The span lengths vary between 33 and 39 m. The viaducts are made up of steel box girders with a reinforced concrete deck slab connected every 3 m by transverse cross girders. There are roller or rocker bearings on reinforced concrete portal piers supporting the steel boxes.

Inspections carried out over a period of time had revealed a number of issues with the viaducts substructures and bearings (see Figure 5.16). These were as follows:

- The bearings, especially the roller bearings, were in a poor condition and were not functioning as intended.
- The pier concrete was in poor condition underneath the bearing plinths, and there was spalling and exposed reinforcement. In

04/03/2008

Figure 5.16 Viaduct pier showing roller bearing.

addition, there was inadequate reinforcement underneath the bearing plinths.
- There were large areas of the piers with high levels of chloride contamination.

After a quality/cost tendering exercise, the authority engaged Atkins, as consulting engineers, to design and supervise a scheme of remedial work.

One of the main challenges faced was due to the absolute lack of any provision for the replacement of the bearings in the original design. This meant that the box girders have to be jacked up to allow the removal of the bearings from the piers. In addition, the piers had to be widened and strengthened to allow correct positioning of the temporary jacks. As the existing steel boxes could not support the jacking loads, external bearing stiffeners had to be bolted to the outside of the web directly above jacks, close to the diaphragms.

The final planning permission and listed building consent for the works was granted in February 2010. As the bridge superstructure including the viaducts is a grade A–listed structure, the planning authorities had consulted with Historic Scotland prior to granting approval and consent.

After planning approval, contract documents were prepared and issued to tender. The contract to carry out the remedial works comprised of the replacement of all of the roller bearings and rocker bearings on the steel box girder approach viaducts on the north and south sides. Associated work included the alterations of the pier tops; construction of reinforced concrete corbels at height; permanent strengthening of the viaduct box at jacking points; a temporary jacking system; cathodic protection to the concrete piers; concrete repairs; and all temporary works, including the temporary fixities needed to maintain articulation during replacement. Some box strengthening that was required independent of bearing replacement was also included in the contract.

The form of contract adopted was a quality/cost tender, based on NEC option A, which had been used successfully for the bridge in the past and is used extensively throughout the construction industry in the United Kingdom. Under an NEC, the ethos is based on partnering, although no formal partnering agreement is entered into by the parties to the contract. Under NEC option A, most of the risks identified with the contract are passed to the contractor for pricing at tender stage and an activity schedule that identifies the main elements of work is priced by the tenderers. However, some of the risks in this contract, mainly the extent of the concrete repairs required and the geometry of the box girders for fitting out of the stiffeners, were retained by the authority.

Although the above form of contract was adopted to minimize the difference between the final cost and the tender cost, it was recognized that given the nature of the contract, there was likely to be some additional work. In order to take into account any potential increase in cost, for tender assessment purposes only, 20% of the average of the activity schedule tender returns was assumed to be the sum of the notional additional costs for each tender. This was based on the available construction industry data, which indicated that outturn scheme costs are generally 20% above tendered price.

The tenderers submitted, as part of their tender, percentages for this additional work to include for overheads on staff, plant, and equipment. These percentages were applied to the actual audited cost of any additional work carried out during the contract. These percentages varied between tenderers and were included in the tender assessment and award process.

After the evaluation of quality and cost, the tender submitted by Balfour Beatty scored the highest combined mark when quality and cost were aggregated. The tender submitted by Balfour Beatty for the checked and corrected activity schedule was for the sum of £13.61 million and was accepted as the most economically advantageous bid. The work involved jacking up the viaduct box girders about 1.5 mm to allow the removal of the existing bearings and a contract discipline allowed only one pier per continuous length of deck to undergo a bearing replacement at any one time.

It was also planned that the jacking and lowering (dejacking) operations would be undertaken overnight between 9:30 PM and 4:00 AM, when traffic volumes were relatively low. It was expected that the use of overnight contraflow working during these hours would be sufficient to remove load from the viaducts. However, careful monitoring of heavy goods vehicles was carried out to ensure that the jacking and lowering operations could be sufficiently controlled. During the jacking and lowering operations restrictions on some abnormal loads was required and allowance made in the contract to provide an 8-week advance notice of these operations to abnormal vehicle haulers.

Given the nature of the works and their proximity to adjacent housing, there were significant environmental issues to be dealt with, especially with regard to noise, dust, and discharge from hydrodemolition. Given this close proximity of the works to housing, the contract included the requirement that the contractor appoint a member of staff as liaison officer and that prior to the commencement of the works, meetings be held with local residents and the community councils to provide as much information on the works as possible and to establish points of contact for the duration of the works.

The works started on site in July 2010 and were completed in October 2012. The final cost of the works was £15.34 million, about 12.7% above the tender price.

5.2.10 Cable band bolt replacement

The main cables of the FRB are its primary load-carrying members and the deck of the suspended spans are linked to the main cables by 192 sets of steel wire rope hangers. These hangers vary in length from 2.5 m at mid-span to 90 m adjacent to the main towers.

The hangers loop around a metal casting, which is through bolted to the cables by a 39 mm diameter waisted shank and 665 mm long high–tensile strength bolts that are pretensioned to an approximately 80-ton load. There are 944 of these bolt assemblies, which comprise the bolts and accompanying nuts and washers.

All of the original cable band bolt assemblies were replaced as part of the hanger replacement contract, which was substantially completed in January 2001.

During an inspection in October 2007, cracking was found in a nut forming one of the bolt assemblies. Fortunately, a mild steel end cap had been installed at each end of the bolts to protect the exposed threads. The nut had clearly failed; however, it was considered that as the end caps were still intact, there was still load remaining in the bolt.

It was not possible to quantify the load remaining in a bolt when a nut had failed. The capacity of the end caps was a concern and this concern was increased by the short length of thread engagement at one end. If cable band bolts lose tension, the clamping action they provide is reduced and the FOS against slippage of the cable band is also reduced. If a hanger is compromised by slippage of a cable band, then an immediate restriction to traffic in the adjacent carriageway would be required. It was realized that work to reinstate a slipped cable band and retension the hangers would take considerable time to complete.

Further cracks in nine other nuts were found, and all 10 bolts were replaced between 2007 and 2009, utilizing the cable gantries to provide access to main cable dehumidification work on the bridge. All of the replacement bolts were taken from a limited supply held in FRB's stores. These had been provided as spares from the hanger replacement contract. Unfortunately, a nut cracked in one of the replacement bolt assemblies about 7 months after installation. It was replaced and no further cracking was recorded. The authority commissioned AECOM to investigate the reasons for the failures. AECOM's report highlighted various issues, including the grade of steel and dimensions of the nuts. There was also some comment on the possibility of the method of bolt tensioning contributing to the failures. AECOM concluded that all 1888 nuts should be replaced in the short to medium term and that the existing bolts should be reused if possible.

FETA made an allowance of £530,000 for this work to be carried out over 2 years between 2013 and 2015 using in-house labor resources. The funding was to cover specialized access platforms and other plant and equipment. However, in February and early March of 2012, a cable inspection carried out by FRB staff revealed 16 more cracked nuts. Of major concern was that on two separate cable bands, the nuts on both bolts were cracked.

Given the above, more frequent inspections and measurements were initiated, for movement, at the two most critical cable bands. In addition, emergency work to replace the defective bolt assemblies at these two cable bands commenced. Following discussions with AECOM, it was concluded that there was a high risk of rate of failure of these nuts in the future. In addition, it was recommend that given the uncertainty around the integrity of all the nuts and the risk that the first proposal to only replace the nuts might not be effective, the decision should be taken to replace the bolts as well as the nuts in all cable bands. This would effectively mean replacing all 944 bolt assemblies.

As a consequence of their background knowledge, the authority appointed AECOM to design and specify the new bolt assemblies, to prepare contract documents, and to supervise the works on site. Flint & Neill was appointed for an independent review of the new bolt assemblies' design.

To ensure that there would be no slippage of the cable bands during bolt replacement, it was decided to fit temporary cable bands beside and abutting the cable bands on the steepest part of the cables. Unfortunately, this meant removing the main cable dehumidification wrapping and sealant locally at each cable band, where a temporary cable band was to be fitted, and restoring the wrapping and sealant after the bolts were replaced. It was not expected that this would cause significant long-term damage to the dehumidification system but would increase the cost of the work.

It was also realized that if the normal procurement process of notification and tendering was followed, commencement of the work would be delayed until summer 2013 and the risk of further cracking within these nuts would increase. If this happened during winter, access to carry out bolt replacement would likely be restricted by either the available daylight or high winds. This could lead to closure of the lane adjacent to the affected cable band for a significant period. If this scenario were to occur on both carriageways at the same time, the disruption to traffic caused by these unplanned restrictions would, from experience, cause unprecedented disruption on both sides of the Forth. As a result, the authority made the decision to procure the works using the negotiated procedure, which allowed negotiations to take place with a single contractor.

Preliminary discussions took place with the contractor, C Spencer Ltd., in March 2012, to determine if they had the capacity to carry out this work and whether they could do so at short notice. C Spencer Ltd. was chosen due to their considerable experience of cable work, not only on Forth but also at Severn and Humber, and they had cable gantries that could be readily modified for this work. They also had a good record of carrying out this type of work within budget. As C Spencer Ltd. installed the dehumidification system on the cables, there would be an advantage if they were responsible for the modifications needed to the dehumidification wrapping and sealant being carried out as part of this work. They confirmed a willingness to carry out the work for the authority.

The terms and conditions of any negotiated contract have to be carefully determined as do the prices put forward by the contractor, as the element of competition has been removed. However, there was a considerable body of pricing information that had been built up, not only at Forth but also at Severn and Humber, relating to working on main cables. This increased confidence that any costs put forward by C Spencer Ltd. could be readily compared to all the other cable works carried out previously on other major United Kingdom suspension bridges.

The design, specification manufacturing, and testing of the new cable band bolts assemblies, which were custom-made items and had to be manufactured from scratch, took about 6 weeks longer than anticipated. Therefore, the first bolts were not replaced until November 2012. However, work progressed well and the contractor employed four cable gantries to access the cable on four fronts. As a result, all 944 bolt assemblies were replaced by September 2013.

5.3 NEW BRIDGE ACROSS THE FORTH AT QUEENSFERRY

In 2008, there were a number of issues relating to FRB that were causing concerns not only to the authority but also to the government and the wider public. These were the condition of the main cables and the likely disruption to users during any replacement or augmentation of the cables; the unknown condition of the anchorages; and the major deck resurfacing and joint replacement work being scheduled.

As a consequence of these issues, the Scottish government decided to commence the process to build a new crossing across the Forth at Queensferry to ensure resilience in the transport system on the east side of Scotland.

ACKNOWLEDGMENT

Working to maintain and operate the FRB in a safe manner to allow users a safe and reliable crossing requires a large number of skilled people to work together effectively. This chapter is dedicated to all those who have been involved in this difficult but hugely rewarding task in the past and at present.

REFERENCES

1. Anderson, J.K., Hamilton, J.A.K., Henderson, W., McNeil, J.S., Roberts, G., Smith, H.S., and Smith, H.S. "Forth Road Bridge," *Proceedings of the Institution of Civil Engineers*, Volume 32, Issue 3, November 1965, pages 321–512.
2. Andrew, A.A.S., and Colford, B.R. "Forth Road Bridge—Maintenance Challenges," Fifth International Cable Supported Bridge Operators Conference, New York, 2006.
3. Mayrbaurl, R., and Camo, S. NCHRP Report 534 Guidelines for Inspection and Strength Evaluation of Suspension Bridge Parallel Wire Cables, Transportation Research Board, Washington, DC, United Sates, 2004.
4. Colford, B.R., and Cocksedge, C. "Forth Road Bridge—First Internal Inspection, Strength Evaluation, Acoustic Monitoring and Dehumidification of the Main Cables," Fifth International Cable Supported Bridge Operators Conference, New York, 2006.
5. Chavel, B.W., and Leshko, B.J. *Primer for the Inspection and Strength Evaluation of Suspension Bridge Cables*, U.S. Department of Transportation, Federal Highway Administration Publication FHWA-IF-11-045, May 2012.
6. Mahmoud, K.M. BTC Method for Evaluation of Remaining Strength and Service Life of Bridge Cables. Report C-07-11. Submitted to the New York State Department of Transportation, Albany, NY, 2011.

Chapter 6

Bronx–Whitestone Bridge

Justine Lorentzson, Dora Paskova, Mary Hedge,
Jeremy (Zhichao) Zhang, and Mohammad Qasim

CONTENTS

6.1 HISTORY

In the 1930s, Robert Moses, chairman of the Metropolitan Council on Parks, envisioned a new highway network for the five boroughs of New York City comprising a series of limited-access arterial routes that he named the Belt Parkway. Moses's vision was that a parkway that circumscribed the borders of Brooklyn and Queens would promote regional mobility. Borough residents would travel only a short distance on local city streets to this new parkway system that would provide direct connectivity to other boroughs.

Around the same period, the Regional Plan Association called for a bridge between the Bronx and Queens, to allow drivers from upstate New York and New England to reach Queens and Long Island. Moses readily supported this idea as it enhanced the intrinsic value of the Belt Parkway system to regional mobility. In 1937, Moses persuaded the New York State legislature to approve the bridge. By this time Moses had become chairman of the Triborough Bridge and Tunnel Authority (TBTA).

After the approval of the bridge by the legislature, the planned 1939 World's Fair in Flushing Meadows, Queens, drove the timetable for construction. Construction began in June 1937 and was completed in record time of 23 months. It opened on the day before the World's Fair—April 29, 1939. As part of the massive bridge project, two major recreational areas were developed: Ferry Point Park in the Bronx and Francis Lewis Park in Queens.

6.2 GROUND BREAKING

Ground breaking was held in June 1937 without a formal ceremony. Instead, Mayor La Guardia and other dignitaries attended the laying of the cornerstone of the Bronx anchorage on November 1, 1937. Some 200 people attended the ceremony at Old Ferry Point. The mayor, Robert Moses, Bronx Borough President James Lyons, and George McLaughlin, financial officer of the TBTA, all dabbed some mortar on the cornerstone.

6.3 DESIGN

Technological advances made it possible for bridges of the 1930s to span longer than ever before. A streamlined look was in vogue at the time, in contrast to older bridges, such as the Brooklyn Bridge, which had massive masonry towers.

Othmar Hermann Ammann (1879–1965), one of the premier bridge designers of the 20th century, was the Chief Engineer for the bridge. The first bridge he worked on was the Queensboro Bridge, followed by a role in the design and construction of the Hell Gate Bridge, as Gustav Lindenthal's assistant. He was also the designer for the George Washington and Bayonne Bridges, both the longest spans of their type, completed simultaneously, under his direction ahead of schedule and under budget. Soon after the completion, he was hired by the TBTA in 1934 as its chief engineer, to take over the planning and construction of the Triborough Bridge.

The Bronx–Whitestone Bridge (BWB) was Ammann's second bridge for the TBTA after the Triborough Bridge (renamed the Robert F. Kennedy Bridge). At the time of its opening, it was the fourth longest suspension bridge in the world. It advanced state-of-the-art bridge construction

Figure 6.1 Side elevation view of the BWB.

significantly. In the 1930s, architects embraced modernism—emphasizing spare, clean lines, without adornment or decoration (Figure 6.1). Amman's design for the BWB, embodying aspects of modernism, was a major departure from past suspension bridges. His use of cellular towers to create a rigid frame structure avoided the need for diagonal bracing. The suspended span superstructure comprised girders and floor beams only, no stiffening truss necessary. Even the anchorages and adjacent approach viaducts also reflect this simplicity. The towers are open and graceful, and the anchorages mirror their simple arch design. The BWB, at its opening, was one of the most beautiful bridges in the world and influenced the design of other major suspension bridges throughout the world for the next 50 years.

Ammann's staff included the engineer of design, Allston Dana, who had worked with him on the George Washington and Triborough Bridges. The architect for the bridge was Aymar Embury II, who had worked with Ammann on the Triborough Bridge. Consultants for the bridge's design were Waddell & Hardesty; Moran, Proctor & Freeman; Madigan-Hyland; Leon Moisseiff; and Charles P. Berkey.

Robert Moses said, "Othmar Amman was at once a mathematician, a forerunner in the industrial revolution and a dreamer in steel. He was a master of suspension and a builder of the most beautiful architecture known to man, a combination of realist and artist rarely found in this highly practical world."

6.4 BRIDGE STATISTICS

At the bridge site the East River narrows to a width of about 3300 ft between shorelines and to 2250 ft between pierhead lines. The total length of the bridge proper is 3770 ft, with a main span of 2300 ft and two side spans of 735 ft each. The length of the structure including approaches is 6812 ft.

The bridge deck provides a clear height of 135 ft above high water at the channel line near the Bronx shore and 137 ft at the center of the span. When constructed, the total width was 74 ft divided into two roadways, each 29 ft wide, separated by a center curb and flanked by two sidewalks. The original superstructure utilized a concrete grid deck supported on sub-floor beams between stringers, a system Ammann devised for suspension bridges and later used throughout the United States.

6.5 BRIDGE CONSTRUCTION

The construction on the Queens and Bronx towers began in June 1937. The depth to solid rock was several times greater on the Queens side and four caissons were sunk by excavating from the surface in open dredging wells. The work on the Bronx side proceeded at a faster rate than on the Queens side, as the depth to the riverbed was much shallower. The towers were 377 ft high and each tower required 3500 tons of steel. The floor steel erection was started at each tower, progressing toward the center of the main span, and then continued from the towers to the anchorages.

6.6 CONSTRUCTION TIMELINE

June 1937	Ground breaking and first contracts awarded
November 1, 1937	Cornerstone of Bronx anchorage laid down
September 14, 1938	Spinning of the cables
February 14, 1939	Closing members of stiffening girders placed in position

6.6.1 Construction statistics

- Main span—2300 ft
- Span including anchorages—3770 ft
- Length from ramp to ramp—6812 ft
- Span width—74 ft
- Roadways—29 ft
- Cable wire—14,800 mi
- Tower height—377 ft
- Tower Steel—3500 tons
- Concrete—200,000 cubic yards
- Rivets—50,000 in each tower

6.7 WIND MITIGATION

After the collapse of the Tacoma Narrows Bridge in 1940, a bridge with similar superstructure characteristics as the BWB, steps were taken to mitigate the potential for wind-induced vibrations and flutter instability. In 1940, cable stay ropes were added to stiffen the bridge for both torsional and vertical motions. In 1945, a heavy stiffening truss was added on top of the main girders on each side of the bridge to further stiffen the suspended span and improve aerodynamic performance. In addition to adding the stiffening truss, the sidewalks were removed and a third travel lane in each direction was added to the bridge, creating the current configuration of six travel lanes that are on average only 10 ft wide with no shoulders.

While these changes improved bridge behavior, further wind performance enhancements were explored, culminating in the addition of a tuned mass damper in 1985. All of these modifications to the bridge have significantly increased the dead load on the main cables. This issue has been one of the recurring design challenges for the BWB—finding an optimal balance between the dead load carried by the cables while maintaining aerodynamic performance. In 2003, wind fairings were installed and the stiffening truss was removed to improve the aerodynamic performance of the bridge with the goal to reduce some of the added dead load accumulated over the life of the bridge. For more details on the implementation of the wind fairing, see Section 6.10.

6.8 INSPECTION

The Federal Highway Administration first established the requirements for nationwide biennial inspections in 1971 [1]. While the inspection program is thorough and comprehensive, with the BWB being a suspension bridge, the MTA Bridges and Tunnels (MTABT) has always directed special attention toward the main cable system including the anchorages and eyebars. Project-specific or condition-driven in-depth inspections are routinely performed outside of the biennial inspection process. The first recorded inspection of the cable bolts and cable bands took place in 1955. Since then, numerous inspections of the main cable and supporting systems have been conducted with increasing frequency (see Table 6.1).

6.8.1 Eyebars

In 1990, the eyebars were inspected and found to have suffered significant section loss that was in part attributed to water infiltration into the anchorages. This precipitated the design of load transfer girders to reduce the load on the eyebars. A prestressed and tapered girder system was chosen to help relieve the load acting on the eyebars. The girder was prestressed against the strand shoes of the worst corroded eyebars using two end anchor rods

Table 6.1 Years and foci of major inspections and studies on the BWB

Year	Focus
1955	Retightening of cable bolts and band
1982	Aerodynamic performance
1987	Tuned mass damper
1990	Biennial, main cable, eyebars
1993	Floor beams and rocker bearings
1994	Biennial, main cable, bridge clearance, cable bolts
1995	Biennial and live loading
1996	Special inspection and hydraulic vulnerability
1997	Biennial and conceptual fendering system
1998	Special inspection
1999	Biennial
2000	Cable inspection, anchorage load cells, special inspection
2001	Biennial, eyebar
2002	Hollow sounding concrete, wind study
2003	Biennial
2004	Main cable, orthotropic deck
2005	Biennial
2007	Biennial
2008	Special inspection, orthotropic deck
2009	Orthotropic deck field testing
2010	Special inspection, weigh-in-motion study
2011	Biennial
2012	Special inspection, post-Sandy inspection
2013	Light pole analysis, Queens approach inspection

and a series of pins and simply supported beams. In addition, a dehumidification system was installed inside the anchorages' eyebar chambers to prevent further section loss. This was the first dehumidification chamber installed on an MTABT bridge.

6.8.2 Main cables

Until 1996, the only inspections conducted of the main cable were visual inspections. Between 1996 and 1998, the Authority commissioned a complete in-depth inspection of the main cable. The entire main cable on both sides was unwrapped and wedged open for inspection and wire sampling (Figures 6.2 and 6.3). Broken wires were repaired, samples of various wires along the length of the cable were taken for strength testing, and the cables were oiled and rewrapped. Based upon the strength of the wire samples, the remaining strength of the cable was estimated. At the completion of this investigation, it was recommended that the dead load carried by the main cables be reduced as much as practical in order to improve the factor of safety

Figure 6.2 Opening cable for inspection at the BWB.

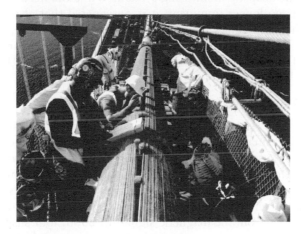

Figure 6.3 Wedging cable for inspection at the BWB.

of the main cables and reduce the rate of deterioration of the main cables. Additional recommendations were made to install an acoustic monitoring system to listen for wire breaks, begin design for cable replacement, and, due to the age and condition of the BWB main cables, to continue to monitor the condition of the cables through partial cable openings every 5 years.

Based upon the recommendations from the original in-depth inspection of the main cables, an acoustic monitoring system was installed in 2001 and projects to remove the stiffening truss and replace the concrete grid decks, as well as a scoping study to determine the feasibility of cable replacement, were initiated. The acoustic monitoring system installed in 2001 was one of the first applications of this technology on a suspension bridge cable.

While the scoping study was being performed for cable replacement, a few cable panels were reopened and reassessed in 2003 to determine the rate of deterioration. It was recommended that several panels on the cables should be reopened and reassessed every 5 years until cable replacement took place. In 2008, the most recent cable investigation was performed. By the time this investigation was performed, the stiffening trusses had been removed and the concrete deck had been replaced with a lighter orthotropic deck. Four panels on each cable were opened, carefully selected based upon the results of the initial full cable opening, as well as the number of potential wire breaks identified by the acoustic monitoring system.

The results from the 2008 cable investigation and subsequent modeling, coupled with the acoustic monitoring, indicate that the rate of deterioration of the cable wires has slowed, due to the oiling and rewrapping of the main cables in 1998 together with the reduction in the dead load on the cables as a result of redecking of the suspended spans and removal of the stiffening truss. Since the reduction in dead load was completed, virtually no additional wire breaks along the main cable outside of the anchorages have been recorded by the acoustic monitoring system. While the need to replace the main cables is no longer imminent, it is anticipated that at some point in the future, the main cables may need replacement. Therefore, a scheduled inspection of the main cables approximately once every 5 years is still recommended. The next cable inspection will be performed in 2016.

6.8.3 Suspender ropes

In 2003, a total of eight suspender ropes were removed for testing to determine the actual remaining strength of the suspender ropes and the interior condition of the suspender ropes and their sockets. In 2006, nondestructive testing was performed on the majority of the suspender ropes by using magnostriction, a nondestructive testing technique that determines section loss inside the rope by sending magnetic pulses up the length of the rope and reading the reflections from the magnetic signal. In 2008, four suspender ropes were removed for destructive testing, which corroborated the findings of the magnostriction testing. Additional suspender ropes will be tested in 2016, to determine when replacement of the suspender ropes should be undertaken.

6.9 MAINTENANCE

Regular maintenance of any bridge is necessary to prolong its life span and keep it in a state of good repair. Suspension bridges are particularly challenging bridges from a maintenance perspective, given the inherent difficulties of gaining access to, inspecting, and rehabilitating elements such as main cables, suspender ropes, and eyebars.

Regular cleaning and painting is one of the most effective means of increasing the life span of a suspension bridge. All bridges owned by the MTABT are subject to a cyclic painting schedule. Most recently, the suspended span was blast cleaned and painted between 2006 and 2007. In 2009, the exterior of the steel towers were fully blast cleaned and painted, both above and below the roadway level. The main cables and suspender ropes were also overcoated to minimize water infiltration into the cables and suspender ropes. They are scheduled for another overcoat during the 2015–2019 capital program.

In 2013, the anchorage dehumidification was replaced with a much more efficient system in order to ensure that the eyebar chambers are kept at the necessary low humidity levels to eliminate corrosion of the eyebar and cable strands in the anchorages.

Main cable dehumidification is another option that is being considered for the future to enhance the service life of the main cables. While anchorage dehumidification chambers are relatively common in the United States, main cable dehumidification is a relatively new phenomenon in the United States, although it is widely used internationally. The technology is based on injecting dry air along the main cable. This helps ensure that any moisture on the main cable is removed. Main cable dehumidification is already in use in suspension bridges in Japan, United Kingdom, Sweden, and in some U.S. bridges. MTABT is funding research on the effectiveness of this technology on existing oiled cables that represent the cable conditions on two of MTABT bridges, including the BWB.

MTABT has created an internal working group, the Cable Working Group, with the official mission of developing a "focused, systematic and consistent approach to preserve B&T's bridge main cables, suspender ropes and anchorages and make clear and concise policy recommendations to the Chief Engineer that will maximize the service life and further the preservation and long-term health of these most critical bridge elements." The group comprises personnel from both design, bridge painting, research, and inspection units, along with facility engineering staff. This ensures that there is a balanced outlook, with the group being aware of the latest technological innovations in this area and of the needs of MTABT specifically.

A set of recommendations that focused on a general inspection program for the main cables, anchorages, and eyebars was issued by the group in 2011. The prominent features included a 5-year oiling cycle for the cable strands and eyebar inspection of all suspension bridges and a 10-year sealing cycle for the anchorages. The purpose is to prevent and mitigate the inescapable issue of corrosion due to water infiltration. This regular and preventative inspection program will help better preserve the structural integrity of the cables.

6.10 REHABILITATION

The early major rehabilitation projects completed on the BWB discussed earlier, such as the installation of the stiffening truss, were meant to improve the aerodynamic performance of the bridge, while the recent rehabilitation projects completed in the 2000s have been aimed at decreasing the dead load on the cables, as recommended under the first cable inspection in 1998, while retaining the aerodynamic improvements. This has been the driving force behind the work performed on the suspended spans to date. In addition, planning for the future traffic requirements is another strong driver in project selection for the BWB, greatly influencing rehabilitation projects on the approaches.

In response to the recommendations issued from the 1998 cable inspection, two projects were initiated to reduce the dead load on the main cables. In 2003, the stiffening trusses were removed from the bridge and a triangular fiber-reinforced polymer (FRP) wind fairings (Figure 6.4) were installed at the fascias of the suspended spans. In addition to the structural benefits, namely, reducing the dead load on the cables and improving aerodynamic performance, this restored the original sleek lines of the bridge. These wind fairings were an innovative solution to resolving the dual problem of reducing dead load while enhancing aerodynamic performance, and were the culmination of months of wind tunnel section model studies to arrive at the proper geometry. The fairings are FRP composites that comprise glass fiber skin surrounding a polyurethane core. Since the composite fairings were installed at the stiffening girder level, rather than on top of it, they restored not only the original aesthetic appearance of the bridge but also the unobstructed view for motorists.

More recently, in 2007, the original concrete deck was replaced with a lighter steel orthotropic deck (Figure 6.5), which reduced the dead load on the cables by 20%. The deck is projected to last a minimum of 75 years with proper maintenance. The orthotropic deck was designed to be a part of the lateral system of the bridge to ensure that the bridge performs adequately during wind events. At the time of the replacement of the deck, a new fire standpipe was also installed on the main span in consultation with the New York City Fire Department.

Figure 6.4 Side elevation view of the wind fairings.

Figure 6.5 Installing an orthotropic deck panel at the BWB.

The approach ramps to the BWB, from both the Queens and Bronx sides, have been completely replaced with wider structurally redundant structures to meet current geometric standards by introducing 12 ft lane widths with shoulders instead of the then standard narrow lanes designed at the time of construction. These new structures have been designed to tie into a future upgraded suspended span, with room for expansion to four traffic lanes in the future, depending upon traffic demand.

Major projects such as the ones described above require lane closures and impact traffic, sometimes over the course of years. One of the most challenging aspects of any future large-scale rehabilitation program at the BWB is the coordination of the work with work simultaneously ongoing at its sister bridge, the Throgs Neck Bridge (TNB). The TNB and the BWB serve the same traffic corridor; therefore, traffic shifts between the two bridges depending upon where work that impacts traffic is being performed. The rehabilitation needs of both bridges must be balanced with the needs of the traveling public. Any delays in construction on one bridge will have cascading effects on the other. Therefore, the future rehabilitation plans for both bridges are coordinated for successful completion of these projects with minimal disruption to traffic.

6.11 EVALUATION OF MAIN CABLES

As mentioned earlier, several inspections of the main cables have been performed. When the original cable investigation (BW-10) was performed in 1998, NCHRP Report 534, "Guidelines for Inspection and Strength Evaluation of Suspension Bridge Parallel Bridge Cables" [2], was being developed and had not been finalized for use on the project. However, the consultant performing the inspection of the BWB cables was instrumental in the development of the NCHRP Report 534 method; therefore, an early version of this method was used to evaluate the cable strength.

In 2003, two cable panels were reopened and reassessed following the NCHRP Report 534 method. In 2008, eight cable panels were opened and reassessed. The cables were inspected in accordance with the requirements of NCHRP Report 534 [2]. The strength calculations and remaining life projections were performed using two independent methods. The first method was as detailed in NCHRP Report 534 and the second method was a proprietary method developed by Bridge Technology Consulting [3].

A total of 144 wire samples, approximately 15 ft in length, were removed from the cable as part of the field inspection. Samples were selected prior to cable opening by using a stratified random sampling plan generated by the proprietary method. The term *stratified* is used since the sampling plan was limited to wires within the first 10 rings (approximately 2 in from the surface), due to restrictions imposed by the equipment used to splice and tension the wires. This method of selecting wires is a departure from the NCHRP guidelines, which recommend selecting wires according to the observed stage of corrosion. That being said, the larger number of samples required using the proprietary sampling plan satisfied the number of samples per corrosion stage recommended by NCHRP Report 534.

The sample wires were cut into 621 "standard" test specimens, 18 in in length, and 166 "long" specimens, ranging from 66 to 72 in in length (an additional 59 nonstandard specimens, less than 18 in in length, were not used). The standard specimens were sent to the Carleton Laboratory at Columbia University, where the specimens were tested to determine their ultimate tensile strength and ultimate strain. The long specimens were tested to determine the presence of preexisting cracks in the wires. The fracture surfaces of all specimens were visually inspected under a stereoscopic microscope to check for preexisting cracks. The results of the evaluation of the cables strength using the laboratory test data in conjunction with the NCHRP and proprietary method were relatively close to each other.

6.12 BEST PRACTICES

The key to maintaining and extending the life span of a suspension bridge, such as the BWB, is an integrated plan that encompasses all aspects of inspection, maintenance, and rehabilitation. On the whole, a regular inspection program is essential to collecting the data necessary to analyze the state of bridge components and identify priorities. A maintenance program is then designed around the needs identified by the inspection program. If necessary, major rehabilitation work may need to be performed to upgrade the bridge if maintenance is no longer feasible due to deterioration or lack of cost-effectiveness. Major rehabilitation may also be necessary on the basis of new code requirements or traffic demands.

Institutional knowledge played a key role over the years as the bridge underwent major changes. The continued presence of engineers and project managers who were intimately familiar with all aspects of the bridge was a direct factor in the successful completion of major rehabilitation/replacement projects. This was critical when giving direction to design projects and analyzing results. Our dedicated employees have helped guide the bridge through the numerous transitions necessary to meet the challenge of transporting approximately 40 million vehicles annually.

The establishment of a long-term capital program planning mechanism ensures that even the most complicated of projects were prepared for, from both engineering and financial aspects. Projects are planned 20 years into the future, which mean that the groundwork started well ahead of time. For example, after the need for a new bridge deck was established, the deck-replacement project was put into planning. The project was designed, reviewed, and refined years before it needed to be practically implemented and was completed successfully.

6.13 FUTURE PLANS

Planning for the future at the BWB must meet several goals. First and foremost, the structure must be maintained in a state of good repair. A secondary goal is to upgrade the structure to meet current geometric criteria and build in additional traffic capacity as deteriorated portions of the structure are reconstructed. Given the current width constraints on the suspended span due to the configuration of the bridge towers, anchorages, and cables, two possibilities were identified and studied for improving the geometric configuration and increasing the traffic capacity of the BWB: adding a second level and widening the bridge, both of which are discussed in detail in the following. Widening of the bridge was selected as the more cost-effective and constructible alternative should additional traffic capacity be needed in the future.

6.13.1 Double decking

Adding a second level to the bridge (Figure 6.6) requires significant reinforcing of the towers, replacement of the cable, and enlargement of the anchorages, and all construction will be over existing traffic. In addition, to access a two-level bridge, a second level would have to be constructed on the approach structures (once again over traffic), the toll plaza on the north side would need significant reconfiguration, and the Queens interchange on the south side of the bridge would need to be completely reconfigured. Environmental impacts in the East River will be minimal, since the

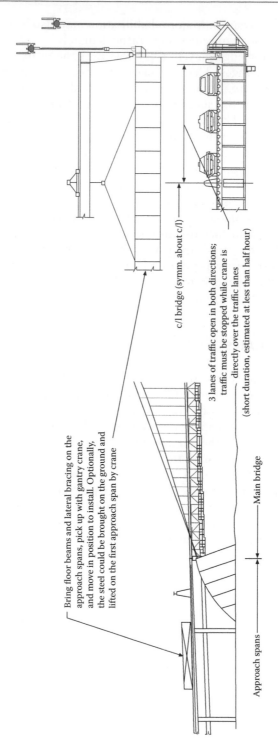

Figure 6.6 Conceptual drawings for a double-decker bridge.

towers are only being reinforced; however, the enlargement of the anchorages may have impacts on the East River on the Queens side. Based upon the complexities discussed above, especially the constructability issues with performing construction over live traffic, along with the significant additional cost, the double-decking alternative was eliminated from further consideration.

6.13.2 Widening alternatives

One alternative is to replace the existing cables with larger cables capable of carrying the widened structure, reinforce the towers, and add additional separated roadways cantilevered outside the existing tower legs (see Figure 6.7). This alternative requires significant reinforcing of the towers and enlargement of the anchorages, similar to the double-deck approach. This alternative minimizes environmental impacts in the East River from the tower work; however, there would still be impacts to the East River from the

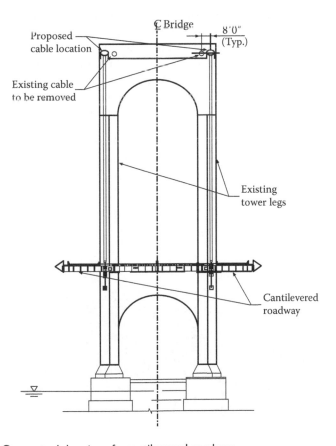

Figure 6.7 Conceptual drawing of a cantilevered roadway.

enlargement of the Queens anchorage. In addition, the newly reconstructed approach viaducts would need significant additional modifications to tie into the separated suspended span roadways. Significant disadvantages associated with this widening alternative are potentially large deflections from heavy trucks on the cantilevered outer roadways and the more difficult operation of the bridge due to the separated roadways.

The other alternative for widening the suspended spans includes the construction of new towers outside of and over the existing towers, with a new pair of cables to support a contiguous roadway with four lanes and shoulders in each direction between the new tower legs. The roadway deck would be widened, and the existing towers removed. The anchorages would be enlarged to a greater extent than under the double deck or widening alternative based on reusing the towers. There will be significant impact to the East River from the construction of the new towers, as well as impacts to the East River from the enlargement of the Queens anchorage. However, this alternative minimizes the need for major modifications to the approach structures. This alternative has several advantages over the double decking and other widening alternative; the old steel towers would be replaced with new towers; roadways will be contiguous, which will allow the operation of the bridge to remain as it is currently; and the bridge will have better aerodynamic performance and would have a better appearance.

Under either widening option, the majority of the construction will have little to no significant impact on the traveling public, the toll plaza would not need to be modified, and the reconfiguration of the Queens interchange is not necessary.

Cable replacement can be conceptualized with the need to modify the suspended span structure to meet current geometric standards and increase traffic capacity. The preferred alternative for accomplishing this

Figure 6.8 Historical aerial view of the BWB.

multipurpose goal is to widen the suspended spans by constructing new towers, widening the anchorages, and extending the floor beams to provide a contiguous four lanes with shoulders in each direction, thus minimizing any modifications needed on the approaches and the toll plaza.

The optimum time for replacing the cables must be carefully planned to avoid premature capital investments, as well as to allow appropriate lead time for related environmental processes, design, and construction procurement. Cable investigations/modeling will continue to be performed approximately every 5 years, coupled with continuous acoustic monitoring, to monitor the remaining cable strength, and the need for cable replacement will be reevaluated based upon the most recent strength modeling results.

The BWB (Figures 6.8 through 6.10) is meeting the needs and requirements of New York today and is well positioned to meet future challenges also. The structure itself might change, but its importance to the region

Figure 6.9 Current aerial view of the BWB.

Figure 6.10 View of the BWB lit up in the evening.

will only increase. The BWB will remain an integral part of the region's transportation network for the foreseeable future.

ACKNOWLEDGMENTS

The authors acknowledge Joseph Keane, Aris Stathopoulos, Chris Saladino, and Michael Bronfman of MTA Bridges and Tunnels for their review comments on this chapter.

REFERENCES

1. Federal Highway Administration. National Bridge Inspection Standards, 23 CFR Part 650, Federal Highway Administration, U.S. Department of Transportation, Federal Register, Vol. 69, No. 239, 2004.
2. Mayrbaurl, R., and Camo, S. NCHRP Report 534, Guidelines for Inspection and Strength Evaluation of Suspension Bridge Parallel Wire Cables, Transportation Research Board, Washington, DC, United States, 2004.
3. Mahmoud, K.M. BTC Method for Evaluation of Remaining Strength and Service Life of Bridge Cables. Report C-07-11. Submitted to the New York State Department of Transportation, Albany, NY, 2011.

Chapter 7

George Washington Bridge

Stewart Sloan and Judson Wible

CONTENTS

7.1 HISTORY

A remote glacial age was responsible for the canyon on our eastern seaboard through which the Hudson River flows. It has been man who, through historic processes, has set up the artificial barrier of two state governments on each side of the river. It is also man who, by his inventive genius has made it possible to span and tunnel the act of nature.

New York Governor Franklin Delano Roosevelt
to the bankers financing the Hudson River Bridge

. . . what is most important, in my opinion, is the linking together of two great commonwealths in a great undertaking which will prove of so much benefit not alone to the section contiguous thereto but to the nation as well.

New Jersey Governor Morgan Larson
to the bankers financing the Hudson River Bridge

The George Washington Bridge (GWB) is an iconic structure with a rich history that is engrained in the development of the region and the authority that brought it to life. The center-span length was nearly double previously attempted span lengths at the time and was the first bridge crossing from New Jersey into Manhattan. Othmar Ammann, the chief engineer of the bridge, worked for the Port of New York Authority in bringing this grand structure from design in 1927 to opening in 1931. The Port of New York Authority is now known as the Port Authority of New York and New Jersey (referred herein as the port authority). Today the GWB carries 14 lanes of traffic across the Hudson River and is considered one of the busiest bridges in the world.

In the port authority's charter ratified on August 30, 1921, the port authority could "improve commerce and trade" anywhere within the 1500 mi^2 of the port district. This broad mandate has helped define their success in driving the region forward with grand projects that benefitted the public and helped grow New York City and the region. In a little over 10 years' time, the port authority took over the operations of the Hudson River tunnel from the company that built it, built one tube of the Lincoln Tunnel, and built four bridges, with the Outerbridge Crossing and the Goethals Bridge built first and the Bayonne Bridge and the GWB coming 3 years thereafter. All this happened through the tumultuous time known as the roaring '20s, which was followed by the Great Depression. These public works project were vital to the continued success of the region.

In 1921, a New Yorker commuter or commuters from New England had several options of getting their car into Manhattan over bridges or through tunnels. However, New Jersey commuters could get their car into Manhattan only via ferry. At the time there were 17 ferries operating between the Battery and Tarrytown.[1] The port authority conducted an extensive study of the amount of vehicles (horse drawn or motor propelled) utilizing the ferry and felt that traffic was going to expand significantly, so getting cars across the Hudson River was a top priority. Analysis concluded that predicted toll revenue from 50 cents per vehicle could meet the construction charges and, therefore, the project was economically sound.[1]

When initially proposed and as envisioned by renowned bridge engineer Gustav Lindenthal, the design for the GWB was a massive structure clad in granite located at midtown Manhattan. Figure 7.1 shows the massive 235 ft wide cross section that Lindenthal proposed with octagonal towers which was supported by eyebar suspension lines. His bridge was estimated to be in the neighborhood of a cost of $180 million to build (very very expensive at the time),[2] but the difficulty also lay in getting the bridge into the connections in the very crowded Manhattan midtown area and the political nightmare that endeavor faced. Lindenthal had a younger engineer working with him named Othmar Ammann, who thought differently about the project. Ammann disagreed with his boss not only in location but also in scale of bridge that should be built.[3]

Ammann's vision for the bridge was for a suspension bridge with a longer span than Lindenthal's but much cheaper to build and at a better location politically. Figure 7.2 shows the cross section that Ammann designed. His

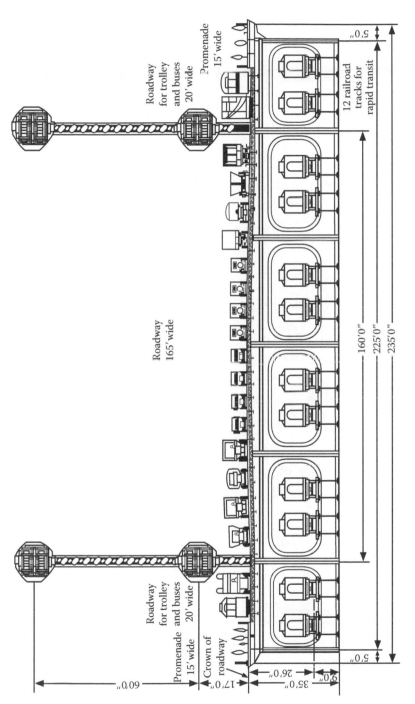

Promenade
15' wide

Roadway
for trolley
and buses
20' wide

5'0"

12 railroad
tracks for
rapid transit

160'0"

225'0"

235'0"

Roadway
165' wide

Roadway
for trolley
and buses
20' wide

Promenade
15' wide

Crown of
roadway

17'0"

26'0"

35'0"

9'0"

5'0"

60'0"

Figure 7.1 A typical cross section of the Lindenthal proposed bridge, eyebar suspension bridge design submitted in 1923. (From the American Society of Civil Engineers, Reston, Virginia. With permission.)

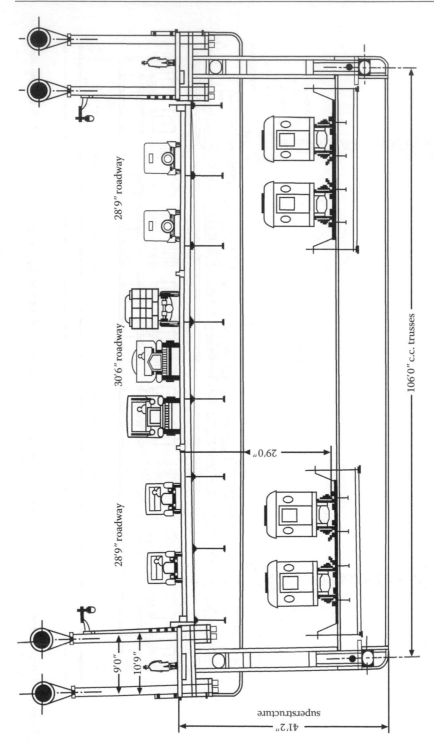

28'9" roadway

30'6" roadway

28'9" roadway

9'0"

10'9"

29'0"

106'0" c.c. trusses

superstructure

41'2"

Figure 7.2 GWB cross section as designed by Othmar Ammann. (From the American Society of Civil Engineers, Reston, Virginia. With permission.)

idea was to build the bridge up near 178th Street, where the face to face of the rock was fairly close, which would help the design. He also proposed to use the weight of the bridge and cables as a damper for wind forces, which was unique for the time.

Ammann's leaving Lindenthal's firm and joining the port authority in 1925 would enable Ammann to design all of the bridges from New Jersey into New York including the four mentioned above. As it turns out, the up-and-coming engineer Ammann would surpass Lindenthal in publicly recognizable bridges through his work on the Golden Gate Bridge, the Verrazano Narrows Bridge, and the Delaware River Memorial Bridge. The GWB was the feather in his cap in the fact that it was the largest suspension bridge span ever designed, and it was able to be built for roughly $60 million, which was much less than Lindenthal's projected cost. In addition, Lindenthal continued to seek out building his bridge even after the GWB was built, presenting before Congress.[4]

When the Holland Tunnel was opened to vehicular traffic in 1927, commuters were able to get into the city, but vehicular traffic also moved through lower Manhattan to get to New England. This route of travel at times would lead to heavy congestion in the city and enhanced the need for the GWB to be built to divert traffic to north of Manhattan.

7.2 DESIGN

Both the New York and New Jersey sides of the Hudson in the area of 178th Street had very dramatic changes in elevation. The New Jersey Palisades sloped down 280 ft within 500 ft of the shoreline, with the New York side rising 200 ft in Washington Heights only 1000 ft away from the shore.[2] Figure 7.3 shows an elevation of a part of the palisades today. These elevations were

Figure 7.3 A current view of the New Jersey Palisades just north of the bridge. (Courtesy of the Port Authority of New York and New Jersey, New York, New York.)

very useful for building a suspension bridge in that they allowed the balance spans to be much shorter. The New Jersey anchorage was enabled to be built into the hillside by blasting and removing 40 ft of rock down. The New York anchorage was determined to be more economical and aesthetic to build by creating a very large concrete anchorage connected to the bedrock below.

A War Department decision declared that the clear height of the bridge should be 200 ft above the river and it was encouraged to keep the piers out of the heavily traveled waterway below. This created a main span of 3500 ft with two side spans of 700 ft. This massive structural design was larger than ever attempted before but a nearly perfect fit for the topography at the site.[1]

Due to the tumultuous time during construction, prudence was advocated in the construction of the bridge. The cross section as proposed by Ammann was pared back in construction, but the design loads incorporated the eventual final condition which included a lower deck designed for multiple lanes of traffic as well. There was also considerable buffer in the design to account for concrete cladding to be placed on the structure should the change be deemed advantageous and approved by local municipalities. At the time Ammann thought that if needed the lower deck could be utilized for additional

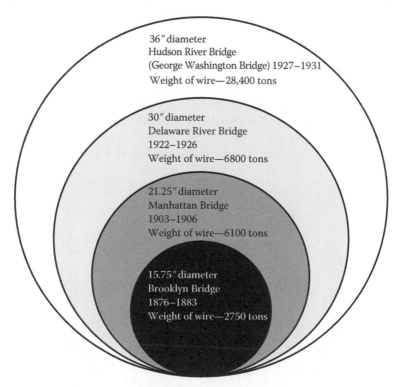

Figure 7.4 Cross-sectional diagram showing a comparison weight of the cables of different suspension bridges at the time. (Courtesy of the Port Authority of New York and New Jersey, New York, New York.)

vehicular lanes. When added later, the additional vertical stiffening trusses of the lower deck also helped combat wind loading. In short, his design in 1929 was bold enough to take into account the addition of the two center lanes in 1936 and the addition of the lower level in 1962 which occurred under his supervision. In general, in the upper and lower roadway condition, the towers support 39,000 lbs per linear foot of the center span, 41,400 lbs per linear foot of the side spans, with live loads, vertical uplifts of the anchorage, and its own weight of 20,000 tons, in short, a massive loading.[2]

Although a common practice now, the design for the GWB cables and suspenders utilized a newer idea that the deflection theory would help lower the weight of the spans and still provide the necessary strength and flexibility to be safe for wind and traffic loading. This was applied successfully to the Manhattan and Delaware River Suspension Bridges previously but not a commonly accepted practice. In addition, the massive size of the cables would help transfer the wind loading to the towers.[2] Figure 7.4 shows a comparison of cable sizes for several suspension bridges.

7.3 COSTS AND SCHEDULE

Building the GWB was a massive undertaking covering nearly four and a half years. The ground breaking was held on September 21, 1927; a photo of the event is shown in Figure 7.5. Progress was made with very little

Figure 7.5 September 21, 1927 groundbreaking ceremony of the GWB. Othmar Ammann is on the far right with the shovel. (Courtesy of the Port Authority of New York and New Jersey, New York, New York.)

interruption with the opening ceremonies held on October 24, 1931, in elaborate events, as shown in Figures 7.6 and 7.7, and officially opening up to vehicular traffic on October 25, 1931. Of note in Figure 7.7, you can see the nonfinished center portion of the deck. The center portion as shown had a temporary deck built and used during the opening ceremonies for spectator standing room. Table 7.1 highlights the contracts that were let

Figure 7.6 Opening ceremonies held on the bridge on October 24, 1931, with Governor Franklin Delano Roosevelt speaking at the grand opening. (Courtesy of the Port Authority of New York and New Jersey, New York, New York.)

Figure 7.7 GWB opening day ceremonies, October 25, 1931. (Courtesy of the Port Authority of New York and New Jersey, New York, New York.)

Table 7.1 Contracts let during the building of the GWB, their award and completion
dates, and the approximate cost per contract

Contract	Description of work	Award date	Completion date	Approximate cost
HRB-2	New Jersey tower foundation	Apr. 15, 1927	May 29, 1928	$1,059,000.00
HRB-3	Excavation for New Jersey anchorage and approach	June 2, 1927	May 25, 1929	$1,150,000.00
HRB-4	New York anchorage and tower foundations	Mar. 8, 1928	Mar. 13, 1929	$1,088,000.00
HRB-5A	Towers and floor steel	Oct. 13, 1927	Jan. 26, 1931	$10,760,000.00
HRB-5B	Cables, suspenders, and anchorage steelwork	Oct. 13, 1927	Oct. 23, 1931	$12,193,000.00
HRB-6	Main approach ramp	Apr. 17, 1930	Oct. 23, 1931	$966,000.00
HRB-7	Demolition and removal of buildings for New York approach	Nov. 21, 1929	Feb. 27, 1930	$150,000.00
HRB-8	Vehicular tunnel in West 178th Street of New York approach	Apr. 17, 1930	Nov. 1, 1931	$2,138,000.00
HRB-9	Riverside Drive connections of New York approach	Sept. 18, 1930	Nov. 23, 1931	$1,250,000.00
HRB-10	New Jersey approach excavation and miscellaneous construction	June 5, 1930	Jan. 8, 1931	$323,000.00
HRB-11	Paving and miscellaneous construction for the New Jersey approach	Mar. 5, 1931	Oct. 27, 1931	$565,000.00
HRB-12	Paving, railings, and miscellaneous construction on main bridge and New York anchorage	May 21, 1931	Oct. 23, 1931	$493,000.00
HRB-13A	Field office building in Fort Lee	May 21, 1931	Dec. 30, 1931	$195,000.00

(Continued)

Table 7.1 (Continued) Contracts let during the building of the GWB, their award and completion dates, and the approximate cost per contract

Contract	Description of work	Award date	Completion date	Approximate cost
HRB-13B	Heating and ventilation equipment for field office building	May 21, 1931	Dec. 20, 1931	$13,000.00
HRB-13C	Electrical installation in field office building	May 21, 1931	Jan. 8, 1932	$5000.00
HRB-13D	Plumbing system for field office building	May 21, 1931	Nov. 11, 1932	$6000.00
HRB-14	Electrical equipment and installation on bridge and approaches	May 28, 1931	Feb. 26, 1932	$92,000.00
HRB-15	Toll buildings	July 27, 1931	Oct. 23, 1931	$155,000.00
HRB-16	Alterations to certain buildings encroaching on Riverside Drive connections	June 25, 1931	Mar. 24, 1932	$80,000.00
HRB-17	Final field painting of towers	Aug. 1, 1931	Dec. 11, 1931	$27,000.00
HRB-18	Floodlight towers, New Jersey plaza	Sept. 10, 1931	Jan. 16, 1932	$41,000.00
	Total of construction contracts let during the project			$32,749,000.00

Source: American Society of Civil Engineers, Reston, Virginia. With permission.

during the construction and their time frames. The contract costs for the total project were tabulated as shown in Table 7.2. Note the difference in price to create the New York connections versus the New Jersey connections; this was due to the density of buildings on the New York side versus the New Jersey side.

While coming in slightly under budget at $60,000,000, the job was also completed 8 months ahead of the scheduled completion date.[2]

Table 7.2 Summary of the costs of construction of the GWB

Bridge and approach construction contracts	$32,749,000
New York highway connections	$15,254,000
New Jersey highway connections	$3,531,000
Interest during construction	$4,543,000
Discount on the bonds	$3,030,000
Cost of the bridge project	$59,107,000

Source: American Society of Civil Engineers, Reston, Virginia. With permission.

7.4 CONSTRUCTION

The New Jersey anchorage was cored out of the rock in a tunneling method in addition to having roughly 40 ft blasted out of the surface of the Palisades traprock formation. This formation was advantageous for suspension bridge design in that it decreased the amount of concrete needed to build the anchorage as well as shortened the approach span length. Figure 7.8 shows a gathering of folks on the anchorage; notice the elevation of the rock outcrops behind the group.

The New York anchorage consisted of a massive amount of concrete at roughly 110,000 cubic yards. For cost savings, the contractor built a concrete-mixing plant and conveyor belt system to get construction materials to a concrete mixer built near the anchorage, as shown in Figure 7.9. This method was determined by the contractor to be the cheapest to build the anchorage.

There are two separate foundations for the New Jersey tower, each installed by cofferdam and tremie concrete down to the bedrock. The foundations were built with tremie concrete from elevation –75.0 ft to the mud line, then from the mud line up to mean high water at elevation +2.0 ft. For the New York tower, solid rock was considerably shallower at elevation –15.0 ft for the south pier and at elevation –5.0 ft for the north pier. All of the foundations were clad in granite facing.

The massive steel towers were built in friendly competition with each other; Figure 7.10 shows both towers being built simultaneously. Cabling

Figure 7.8 A celebration of the completion of the cables. Note the elevation view of the bedrock excavation required for the New Jersey anchorage. (Courtesy of the Port Authority of New York and New Jersey, New York, New York.)

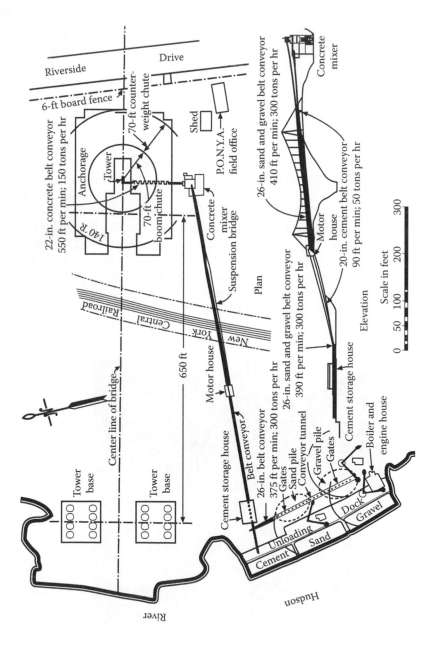

Figure 7.9 A plan view of the concrete mixing operations for the New York anchorage. (From the American Society of Civil Engineers, Reston, Virginia. With permission.)

Figure 7.10 The towers under construction. (Courtesy of the Port Authority of New York and New Jersey, New York, New York.)

operations could not take place until both towers were complete. Figure 7.11 shows the fit up of several sections. Note the number of rivets seen in the photo; there is estimated to be 10 million rivets installed in the bridge. The 40,200 tons of structural steel to build the towers carried them to 560 ft in height. This elevation was supported by 12 columns per tower leg of single-cell box sections consisting of a slight lean from the outside columns and the inside columns being plumb. The cable saddle was placed directly over the inner column of the tower legs, which created an economical advantage by decreasing the size of the floor beam length needed when the cables hung down.

A challenge lay in erecting the main cables across the river. Figure 7.12 shows the cabling being started, as it is dragged across the river on a barge sinking to the bottom of the channel and then raised into place. Figure 7.13 shows the workers creating a worker access platform for building the suspender ropes. The demonstration shown in Figure 7.14 shows a proof

Figure 7.11 Fitting up part of the tower sections. (Courtesy of the Port Authority of New York and New Jersey, New York, New York.)

Figure 7.12 Approaching north side of the New Jersey tower with first cables. Note the severe slope up the Palisades. (Courtesy of the Port Authority of New York and New Jersey, New York, New York.)

concept of how a cable compaction would work. The cables were compacted using 12 thirty-ton jacks as shown. Figure 7.15 shows the compaction apparatus at work. The four main cables are composed of parallel wires carried back and forth across the river, each containing 26,474 galvanized steel wires formed into 3 ft diameter cables that, if laid out end to

Figure 7.13 Building the work platforms for access and suspender cable construction. (Courtesy of the Port Authority of New York and New Jersey, New York, New York.)

Figure 7.14 End of cable section in compressing machine, October 11, 1928. (Courtesy of the Port Authority of New York and New Jersey, New York, New York.)

end, would be roughly 105,000 mi in length or over four times around the earth.[2] In addition, additional measures have been added to protect the cables from the elements.

Once the suspender ropes were in place, a balanced method of installing the floor beams was used to erect the superstructure. An example of a floor beam installation is shown in Figure 7.16. During initial construction, the top chord framing for adding a lower roadway was put in place.

Figure 7.15 GWB workers. (Courtesy of the Port Authority of New York and New Jersey, New York, New York.)

Figure 7.16 The floor beam sections were raised into place and attached to the suspender ropes. (Courtesy of the Port Authority of New York and New Jersey, New York, New York.)

Once the floor beams and stringers were installed, the deck was placed to provide two travel lanes in each direction. The center portion of the deck was added in 1936, and the lower level of the deck was added in 1962. Figure 7.17 shows the first structural sections of the lower level being prepared for installation. Ammann was brought in to consult on the addition of the lower level due to the complexity involved with balancing the loads. Figure 7.18 shows the underside of the completed lower level along with a view of the Little Red Lighthouse that is on the New York side near the towers. The lighthouse was built to originally prevent ships from crashing into the New York tower as the Hudson River is a major waterway.

When buses started using the bridge on a regular basis, commuters would oftentimes ride the bus across the bridge and then disembark on the other side to catch a connection, as shown in Figure 7.19. This led the port authority to create the GWB Bus Station, which serves thousands of commuters every day; it is shown under construction in Figure 7.20. The bus station is managed by the port authority staff that oversee the bridge. In the overview of the bridge, shown in Figure 7.21, you just make out on the opposite side of the bridge the iconic Nervi truss that was used for the roof of the bus station.

Figure 7.17 GWB raising of the first structural steel section for the lower level of the GWB. (Courtesy of the Port Authority of New York and New Jersey, New York, New York.)

Figure 7.18 A view of the Little Red Lighthouse below the bridge on the New York side showing the completed lower level. (Courtesy of the Port Authority of New York and New Jersey, New York, New York.)

Figure 7.19 Buses dropping off commuters on the New York side of the bridge. (Courtesy of the Port Authority of New York and New Jersey, New York, New York.)

Figure 7.20 Building of the bus station to the east of the GWB in New York. (Courtesy of the Port Authority of New York and New Jersey, New York, New York.)

Figure 7.21 Overview of how the GWB looks today. (Courtesy of the Port Authority of New York and New Jersey, New York, New York.)

7.5 INSPECTION

Biennial inspection contracts are issued for contractors to inspect separate portions of the GWB. The contracts for the main-span upper level, the main-span lower level, the mechanized flag hoist, the New Jersey approach roadways, the New York approach roadways, the Trans-Manhattan Expressway structures, and the bus station and bus parking level are all inspected every 2 years. The other structures are inspected every 4 to 5 years.

Most of the structures do not require specialized equipment for inspection; the typical snooper or vertical lift can be used and inspect most of the structures.

Specialized inspection equipment or methods are used on the lower level and the towers. The lower level utilizes four travelers suspended below the lower-level structure for access during inspection. They are also used for repair contracts depending on the type of repair. They are suspended from beams connected to the floor beams specifically built for the travelers. The traveler "car" utilizes an electric motor to drive the wheels. The upper level is inspected from above and below by lane closures. The towers are inspected by several methods, where possible inspection is from the vertical elevator, but a majority of the inspection of the tower is either by tethered climbing or through access hatch covers in the steel or, if required inspection of the tower or tower components, can be from temporary installed access platforms.

7.6 MAINTENANCE

The GWB has a team of workers that act as caretakers of the Bridge. The Safety, Engineering, Maintenance, and Construction (SEMAC) team makes repairs, improves bridge safety, monitors the bridge, and corresponds directly with the bridge police force as well as with the port authority design staff when issues arise. They are also responsible for localized painting of the bridge. In addition, the SEMAC staff comes up with repair concepts that are workable in the field and are easy to maintain.

7.7 FUTURE PLANS

The GWB is over 80 years old and has held up very well, thanks to the maintenance schedules and efforts of the port authority and rehabilitation contracts. Current plans for the GWB include the replacement of the slightly less than 600 suspender ropes. In addition, approach roadways and ramps are projected to be repaired in the next 10 years. The port authority realizes that these maintenance measures need to be undertaken to ensure a long and healthy life for the GWB and the surrounding ramps.

The port authority is also looking at erecting a higher sidewalk fence to protect pedestrians. This addition is being evaluated for wind loading on the structure and visual impact for local municipalities.

As mentioned in the companion book,[5] there are also anchorage-related repairs, including passive drainage and replacement of the sump pumps, as well as sump access platforms. Dehumidification is ongoing in the anchorage.

7.8 CONCLUDING REMARKS

The GWB is ingrained in the fabric and bustle of city life. The GWB has been a landmark bridge for New York City and the surrounding communities, providing a key access point to the city and serving as an inspirational structure stretching across the Hudson River. Othmar Ammann's vision for a grand structure standing for years to come led to an advanced design and calculations during the golden age of suspension bridges in the United States. The port authority, acting as caretaker, has continued to maintain, improve, and restore the bridge for millions of commuters that use the bridge every year. Finished in 1931 and standing tall for 80+ years, the GWB remains a testament to the legacy of the port authority.

REFERENCES

1. *Tentative Report of Bridge Engineer on Hudson River Bridge at New York Between Fort Washington and Fort Lee*, February 25, 1926, Pandick Press Inc. Printers, New York, New York.
2. George Washington Bridge Across the Hudson River at New York, NY, *Transactions of the American Society of Civil Engineers*, Vol. 97, with the Port of New York Authority, 1933.
3. Joe Mysak and Judith Schiffen, *Perpetual Motion*, General Publishing Group, 1997.
4. The Hudson River Bridge, November 1929, Pamphlet put out to support financing of Hudson River Bridge bonds, VBA p. 71747, Kissel, Kinnicutt & Co.
5. Alampalli, S., and Moreau, W.J. (Eds.). *Inspection, Evaluation and Maintenance of Suspension Bridges*, CRC Press, 2015.

Chapter 8

Angus L. Macdonald Bridge

Ahsan Chowdhury, Anna Chatzifoti, and Jon Eppell

CONTENTS

In April 1955, the Angus L. Macdonald Bridge (Macdonald Bridge) opened to accommodate the growing communities of Halifax and Dartmouth, in Nova Scotia, Canada (Figure 8.1). Just 15 years later, in July 1970, a second long-span suspension bridge, the A. Murray MacKay Bridge, was opened. The Macdonald Bridge has a main span of 1447 ft (441 m) and side spans of 525 ft (160 m) each. It was built as a two-lane bridge with a narrow sidewalk and converted to three lanes with a pedestrian walkway and bicycle lanes in 1999. It is undergoing replacement of the suspended spans of the bridge in 2015–2017.

The following sections will briefly discuss the history, original design, and construction of the Macdonald Bridge, as well as aspects of the inspection, evaluation, rehabilitation, and maintenance program that ensure its long-term integrity.

8.1 HISTORY

The Macdonald Bridge was built as a two-lane suspension bridge connecting Halifax and Dartmouth over the Halifax Harbour, in Nova Scotia, Canada. Before the bridge was built, traveling between Halifax and Dartmouth was either by a car ferry or by a 12.5 mi (20 km) long drive around Bedford Basin. The Macdonald Bridge was the third bridge to cross the harbor. The Canadian National Railway built the first bridge in 1884. It was a single-track timber trestle bridge with a steel swing section in the middle to allow shipping access to and from the Bedford Basin. It was 40 ft (12 m) wide and 1300 ft (400 m) long. On September 7, 1891, a severe storm blew it down. The bridge was rebuilt in 1892 but it lasted hardly a year. In 1893, on a calm July night, the bridge simply washed away.

Local folklore has it that a British naval officer, Captain John Smith, fell in love with the daughter of a Mi'kmaw chief living on the shores of Bedford Basin. During a secret rendezvous, the couple stole away under the cover of darkness to the officer's ship, which sailed for England the next day. The chief was so distraught and outraged that he placed a curse proclaiming, "Three times will the white man attempt to bridge Chebucto

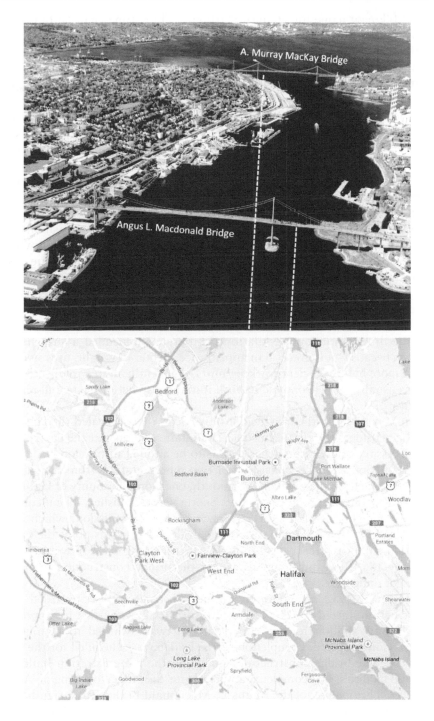

Figure 8.1 Halifax Harbour Bridges: Angus L. Macdonald and A. Murray MacKay Bridge.

(the Mi'kmaw name for Halifax Harbour). Three times will he fail. The first will be a great wind, the second in a great silence, and the third will be with great death" [1].

Although the story supposedly predates any harbor bridges, some believe that it reflects what happened to the first two bridges. At the opening of the Macdonald Bridge in 1955, a Mi'kmaw chief removed the curse. It was not until after World War II that plans were laid for building a third bridge. Postwar Halifax was growing and harbor ferries could not handle the increasing traffic. For the generations still shaking from the Great Depression and World War II, the Macdonald Bridge was to become a symbol of hope: a sign that Nova Scotians could span the tragedy and destruction of the past two decades and build a secure and prosperous future.

In 1948, the province of Nova Scotia, under Premier Angus L. Macdonald, hired Dr. Phillip Louis Pratley to investigate the feasibility of building a bridge across the harbor. Planning for the bridge stalled when a strong campaign was launched favoring a tunnel, arguing that the cost of a tunnel would be less than half the cost of a bridge. Tunnel proponents suggested a three-lane tunnel with eastbound, westbound, and emergency lanes. The government asked Pratley to study the feasibility of a tunnel as well as a harbor bridge. Pratley's study rejected the tunnel as too costly and needing continuous ventilation. Pratley recommended a suspension bridge, because "it is graceful in appearance, lighter in weight, uses steel at much more highly efficient working units, is generally more rapidly erected and is usually more economic in both initial construction costs and maintenance" [1].

On May 6, 1950, the province of Nova Scotia incorporated the Halifax Harbour Bridges (HHB) to construct, maintain, and operate the bridge and the necessary bridge approaches. It was to be a tolled suspension bridge. A board of commissioners consisting of two representatives from the city of Halifax, one from the town of Dartmouth, one from the county of Halifax, and three from the province of Nova Scotia was established. The incorporated legal name is Halifax–Dartmouth Bridge Commission, which was rebranded in 2009 as Halifax Harbour Bridges. From this point the organization will be referenced as HHB. The first meeting of HHB was held on January 18, 1951, and preparations for the bridge construction immediately got underway.

In January 1952, the contracts for the substructure were awarded and construction started in March 1952. In April 1952, a contract for approximately $4.9 million for the superstructure was awarded to Dominion Bridge Company Ltd. Completion of the bridge was scheduled for the end of 1954. The bridge construction, including the administration building and toll facilities, was completed and ready for the official opening on April 2, 1955. The opening of the Angus L. Macdonald Bridge was a significant milestone in the history of Halifax and Dartmouth, but the full impact of the bridge on the area would not be realized for a few years.

The bridge was a more attractive trip than the vehicle ferries that had been in operation since 1876. On March 28, 1955, the vehicle ferries carried 1830 vehicles. Six days after the bridge opened, the number of vehicles plummeted to 224 and within a year the vehicle ferries would cease operation. Around the same time, the passenger ferries reduced their carrying capacity from three 40-passenger boats and one 93-passenger boat to smaller boats that could operate with only a three-member crew.

The opening of the bridge improved the residents' quality of life. People could cross the harbor quickly, eliminating the long lineups and waiting times for the ferry. This eased travel for sporting events, concerts, and other entertainment, and Dartmouth quickly grew as a result. The bridge also allowed for speedy passage of ambulances to hospitals in Halifax. Shopping centers, land development, and residential construction in Dartmouth accelerated within the first 5 years of the bridge's operation. With the increase in population from 20,000 in 1955 to 50,000 in 1961, Dartmouth became a city.

In less than a decade, bridge use grew significantly and the bridge became a bottleneck. Traffic crawled over the span each morning and evening, triggering the need for a second bridge. There was disagreement between the city of Halifax and the city of Dartmouth about the location of a proposed second bridge. The idea of adding a third lane on the Macdonald Bridge was suggested, but it was soon rejected. It was considered impossible to widen the roadway sufficiently for three lanes beyond the existing two-lane roadway at 27 ft (8 m).

When the cities failed to agree on a location, the province of Nova Scotia stepped in and, with the help of HHB, chose a site about 2 mi (3 km) north of the Macdonald Bridge at the harbor narrows, close to the site of the 1918 Halifax explosion. The consulting engineers, Pratley and Dorton, were hired in 1963 to design and supervise the construction of the four-lane suspension bridge with a main span of 1440 ft (439 m) and interchanges requiring 17 overpasses of various types. The A. Murray MacKay Bridge was opened in July 1970. It drew considerable international interest since it incorporated the use of steel orthotropic decks on a suspension bridge for the first time in North America. Notably the bridge also included the use of an aerodynamic study for the construction.

8.2 DESIGN

The Macdonald Bridge was originally designed between 1948 and 1952 by Dr. Phillip Louis Pratley. Dr. Pratley was born in Liverpool, England, in 1884. He was educated at the Universities of Victoria and Liverpool, where he obtained his bachelor of science degree in 1904, bachelor of engineering degree in 1905, master of engineering degree in 1908, and his doctor of engineering title in 1939. Right after graduating in 1905, he immigrated to Montreal and was hired by the Dominion Bridge Company, where he designed numerous railway bridges. He was the chief designer for one of the longest double-cantilever bridges in the world: the 1804 ft (550 m) long railway bridge across the Saint Lawrence River at Quebec. He was a member of the Universal British and North American Esperanto Associations, the Institute of Civil Engineers, the Engineering Institute of Canada, the Institute of Structural Engineers, and the American Society of Civil Engineers. Dr. Pratley died in 1958. At the time of his death, Dr. Pratley was working on the Champlain Bridge in Montreal, Quebec, and the Cornwall North Channel Bridge in Cornwall, Ontario.

The original design of the Macdonald Bridge was for two highway traffic lanes, one sidewalk along the south side of the bridge deck, a duct way along the north side, and a 30 in (760 mm) water main underneath the deck (Figure 8.2).

The overall length of the bridge, from the Dartmouth abutment to the Halifax abutment, is 4418 ft (1347 m). It consists of two approaches, from the abutments to the cable bents; two suspended side spans of 525 ft (160 m) in length each, from the cable bents to the main towers; and one main suspended span, from main tower to main tower, which is 1447 ft (441 m) long. The original roadway was 27 ft (8 m) wide.

The Halifax approach consists of a continuous girder span (comprising three 85 ft [26 m] spans) and a single 233 ft (71 m) truss span. The Halifax approach spans are supported by three concrete piers, plus the Halifax abutment and the cable bent pier (which is made of steel). The Dartmouth approach consists of two continuous girder spans (each comprising three 85 ft [26 m] spans), one continuous girder span (comprising two 150 ft [46 m] spans), a single 124 ft (38 m) truss span, and a continuous truss span

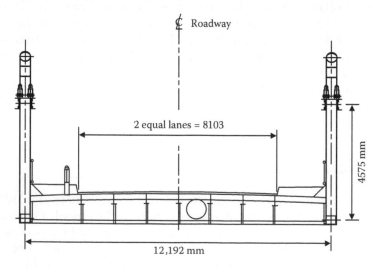

Figure 8.2 Section through suspended span original construction in 1955.

(comprising three 165 ft [50 m] spans). The Dartmouth approach spans are supported by 10 concrete piers, one steel pier, the Dartmouth abutment, and the cable bent pier.

The suspended spans are supported by two main cables through 146 hangers of 1.75 in (44.5 mm) nominal diameter. Each main cable has an outer diameter of 13.5 in (344 mm) and consists of 61 galvanized wire rope strands of 1.478 in (38 mm) nominal diameter, wrapped all together with galvanized wire and painted. Each main cable weighs 475 tons (430 tonnes) and is 3116 ft (950 m) long, spanning from the Halifax anchorage to the Dartmouth anchorage. The main cables enter into an anchor chamber, within the anchorage, and splay out to individual strands. The strands are anchored into a massive 20,000-tons (18,000-tonnes) concrete block.

The deck of the approaches was a 6.5 in (165 mm) thick reinforced concrete slab supported on floor beams and main carrying members, which are either built-up plate girders or trusses. The concrete deck has since been replaced in 1998/1999 and this is discussed in Section 8.7. The deck of the suspended spans is a 3 in (75 mm) steel grid filled with concrete. The grid is 4×4 in^2 (100×100 mm^2) and comprises 0.75 in (19 mm) diameter longitudinal rods welded to transverse lock I sections. At the bottom of the grid there is a 20 gauge (0.9 mm) steel sheet welded to the lock I sections, which provided the formwork for the concrete when it was placed. The grid deck is directly supported by longitudinal stringers at 4 ft (1.3 m) centers. The stringers are connected to the transverse floor beams so that the top flanges are level and the floor beams are connected, in turn, to the lower part of the vertical members in the longitudinal stiffening truss. Along the entire length of the bridge there is horizontal cross bracing to laterally support the bottom flange of the built-up plate girders, the bottom chord of the approach trusses, and the bottom chord of the stiffening trusses.

The most attractive elements of the Macdonald Bridge are the two steel main towers. Each main tower is 302 ft (92 m) tall and weighs 900 tons (820 tonnes). The main components of the tower are the two legs (which transfer the main cable loads to the tower concrete foundation), the tower struts, and the tower diagonals. The struts and diagonals make up the tower-bracing system, which prevents the tower legs from buckling. The deck of the bridge is very close to midheight of the main towers, which provides an attractive form and about 160 ft (49 m) clearance for navigation.

8.3 ORIGINAL CONSTRUCTION

Before the substructure work began, a number of houses in Halifax and Dartmouth had to be removed.

On March 31, 1952, excavation started in Dartmouth. By the end of 1952, the excavation was completed for the main cable anchor blocks, 50% of the concrete had been placed for the Dartmouth anchorage and

25% for the Halifax anchorage. Next, the construction of the caissons (a caisson is a watertight chamber in which underwater excavation can be performed) for the Dartmouth main tower foundation began on land. The caissons were later floated to the designated location, approximately 600 ft (180 m) from the Dartmouth shore, and sunk to the harbor bed. The deepest excavation was 70 ft (21 m) underwater for the Dartmouth main tower foundation. The caissons were later filled with concrete to form solid columns atop which the tower shoes were installed and the towers were bolted. The quantities of concrete poured into the foundations of the Halifax and Dartmouth main towers were 5600 tons (5100 tonnes) and 10,000 tons (9100 tonnes), respectively. Both anchorages and all the bridge piers were completed by February of 1953, requiring 70,000 tons (64,000 tonnes) of concrete.

The fabrication of the superstructure was completed off site during the first 2 years of construction. By the end of 1953, 8 of the 12 spans of the Dartmouth approach had been erected. The two towers were erected between January and May of 1954. The following summer, the catwalks for the main cable construction, main cables, stiffening trusses, and the deck were installed.

Two catwalks were extended, over the top of the two main towers and over the cable bents, from the Dartmouth anchorage to the Halifax anchorage. Along these catwalks, skilled bridge workers placed the main cable strands, later wrapping the main cables and installing the cable bands and hangers. The suspended spans stiffening trusses were next attached to the hangers, in 33 ft (10 m) long pieces, and bolted in place. The procedure started at the center of the bridge and proceeded in both directions. The steel grid deck for the roadway was laid, concrete was placed to the top of the grid, and the bridge structure was painted.

The construction of the Macdonald Bridge lasted for 37 months and required 4400 gallons (20,000 L) of paint, 8000 tons (7300 tonnes) of bridge steel, and 430,000 rivets installed in the field. Five people lost their lives during the bridge construction. On a Saturday afternoon in July 1954, a sudden and severe storm hit the catwalks, throwing a bridge worker, Jean Marie Belanger, to his death in the water below. Four other men faced similar fates during the bridge's construction: Rudolphe LaRocque, Louis Benoit Pelletier, Arthur McKinley, Yvon Moreau, and L. J. McMahon. A plaque on the bridge is dedicated to those who lost their lives in constructing the bridge.

8.4 INSPECTION

HHB operates an effective and proactive inspection program to ensure that the bridges are structurally sound and well maintained to maximize the useful life of the bridge. There are two types of inspections: the annual inspection and investigative inspections.

8.4.1 Annual inspection

Each year, the bridges undergo a rigorous inspection conducted by a team of independent consultants and HHB representatives. The objective of the annual inspection is to observe as many components of the bridge as practical to identify items requiring maintenance, determine the course of actions for the current year, and ensure that items from previous inspections are being addressed properly. Items are categorized as near, medium, or long term. The annual inspection report is used to develop a 3-year maintenance and capital project plan and is also an input to HHB's 20-year plan.

8.4.2 Investigative inspection

Investigative inspections are carried out on particular bridge component(s) or as follow-up on observations made during the annual inspection. The purpose of investigative inspections is to assess the condition, identify the extent of deterioration or damage, determine the root cause and the residual life of the component, and develop an effective repair strategy. Some of the investigative inspections are described in the following.

8.4.2.1 Main cable inspection

The main cable comprises a bundle of 61 galvanized bridge strands in a hexagonal arrangement. The strands are (helical counterclockwise) preformed strands that were individually pulled across the harbor. Each individual wire is galvanized. The 61 strands were compacted into a hexagonal shape, lead paste was applied, shaped cedar wood fillers were installed to give a round shape, and then they were wrapped with an individual galvanized wire and painted.

In 2010, the main cable was internally inspected for the first time since the bridge was built. The inspection was done in two locations: a 32 ft (9.75 m) section near the cable bent on north cable and a 31 ft (9.5 m) section near midspan on south cable. An enclosure was built around the main cable before removing the wire wrapping, cedar wood fillers, and the lead paste (lead powder/linseed oil) coating to expose the outer strands. The exposed strands were inspected and then the cable was wedged to expose the inner surfaces by using ultrahigh–molecular weight (UHMW) plastic wedges (Figure 8.3). The strands of the main cable were found to be in very good condition, with the original galvanizing in place and effective.

8.4.2.2 Cable band bolt tension measurements

The cable bands are cast-steel elements, made in two halves, which are clamped with horizontal bolts to the main cable. The suspended spans are supported by the main cable through 146 hangers. The hangers are

Figure 8.3 UHMW plastic wedge arrangement for inspection of internal strands.

connected to the top chord of the stiffening truss with pins and are continuous over the top of the cable bands on the main cable.

In 2010, an investigation was conducted to determine the tension in the cable band bolts as part of preliminary engineering of the suspended spans' deck-replacement project. Results of the investigation were used in statistical analysis to determine if cable band bolts need to be tightened to avoid cable band slippage during the suspended spans' deck-replacement construction when hanger tension is expected to be much greater than typical in-service condition. A similar assessment was completed in 1996 prior to installation of the third lane on the Macdonald Bridge.

8.4.2.3 Hanger inspection

The nearly 60-year-old hangers are made of 1.75 in (45 mm) diameter wire rope consisting of structural stands (six outer strands and one core strand, each individual wire was hot dipped galvanized) with open spelter sockets at each end. The stiffening truss on the suspended spans is above the road deck and the vertical hangers connect to the top of the stiffening truss. As a result, the hanger connection to the stiffening truss is largely above the salt spray zone from vehicles on the road. The lower end of the hangers, sockets, and connection to the stiffening truss have been inspected and found to be in very good condition. By contrast, on the MacKay Bridge, about 1.5 mi (3 km) up the harbor, the stiffening truss is below deck and the hangers connect to the top of the stiffening truss at road deck level. On the MacKay Bridge there have been numerous hangers with material loss and broken wires, requiring replacement of about 20 hangers.

8.4.2.4 Condition assessment of concrete structures

The Macdonald Bridge piers are mass concrete with little reinforcing steel. The reinforcing is smooth bars and the aggregate is unfractured stones from riverbeds, which was common in 1950s construction. Many of the piers exhibit microcracking and alligator cracks on the surface, and some have more advanced concrete deterioration with spalling and surface loss.

In 2012, the Macdonald Bridge concrete structures were inspected as part of the due diligence for the suspended spans' deck-replacement project. The inspections found that the concrete has suffered from alkali aggregate reactivity (AAR), coupled with cyclic freezing and thawing damage, reducing the strength and quality of the concrete on the outer shell. The AAR has about exhausted. The inner core of the piers remains in good condition. The recommended course of actions was to remove the exterior face of most of the piers and encapsulate with better quality concrete.

8.4.2.5 Road safety assessment

In 2012, HHB conducted a detailed road safety investigation and design for the Macdonald and MacKay Bridges and approaches. As a result of the audit, HHB had to assess the identified deficiencies and prioritize project elements. The project goal was to improve existing roadside design by providing continuity of roadside barriers and end treatments. The project scope included crash attenuator systems, speed management, and signage clarity (positive guidance, clear and concise) for regulatory, warning, and information signage (including way-finding and toll-related signage). The starting point was to prepare an inventory of signs, roadway lighting, pavement markings, light/sign poles, and overhead structures. Structural assessments were conducted where warranted.

8.4.2.6 Inspection of fall-protection components

The requirement for safe access systems, as well as expectations of workers, has evolved significantly since 1955. HHB conducted an assessment of areas accessed at least annually to define whether permanent or temporary access systems were warranted. A priority list has been developed and is being implemented as maintenance painting operations move locations. Permanent horizontal and vertical lifelines and anchor points are considered. Periodic review related to regulatory requirements is conducted and workers are consulted in development of access solutions.

8.5 BRIDGE LOAD EVALUATION

In 2009, a load evaluation was undertaken to provide information about the load-carrying capacity of the bridge and as input to the suspended spans' deck-replacement planning.

The load evaluation of the suspended portion of the Macdonald Bridge involved the investigation of the following bridge components:

- Suspension system
  - Main cables
  - Hangers
- Suspended structure
  - Longitudinal stiffening trusses
  - Transverse floor beams
  - Longitudinal stringers
  - Lateral bracing system
- Towers and cable bents
  - Main legs
  - Diagonal bracing
  - Horizontal struts
- Connectors
  - Wind pins
  - Bearings
  - Some typical structural connections

The evaluation procedure included determining the appropriate bridge loadings, developing a computer model of the bridge and member properties, developing demands from the factored load combinations based on the Canadian Highway Bridge Design Code (CHBDC) requirements, and calculating the member capacities based upon CHBDC formulas. The output was a demand-to-capacity (D/C) ratio for each member and for each loading combination.

A three-dimensional model was created to present the current geometry of the bridge (Figure 8.4) by starting with the original design data and incorporating all known modifications made since the construction of the bridge. In parallel, a survey was conducted of the bridge and the analysis results were compared to survey results. Good agreement was found between the expected geometry and the surveyed geometry.

The survey measured the elevation of the top chord of the truss at each hanger location and the verticality of each leg of the main towers and cable bents. As the geometry of a suspension bridge is sensitive to its temperature, it was important for the survey to be performed when the temperature of the bridge was, as nearly as possible, uniform throughout. These conditions occur only at night, usually near the end of the night, with overcast skies and no wind. It was necessary to have no traffic on the bridge during the survey.

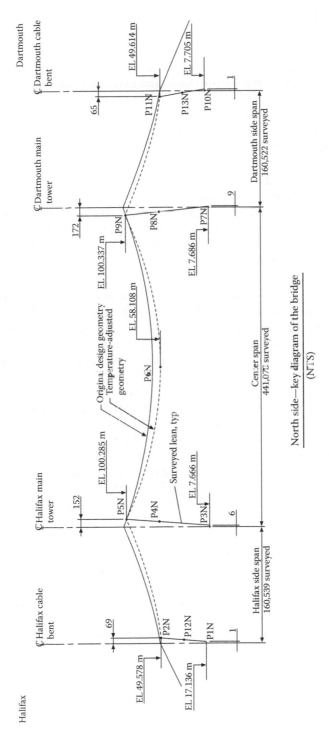

North side—key diagram of the bridge
(NTS)

Figure 8.4 Original design geometry (1955) versus surveyed lean (2012).

The major load changes between the original (1955) state of the bridge and its current state are as follows:

- The superstructure dead load increased by approximately 17% when the modifications during the third lane expansion were made.
- Lateral forces from wind loads used for the load evaluation were similar to those used for the original design. The difference was that vertical wind loads were applied to the suspended structure for the load evaluation, and these were applied eccentric to the centerline of the bridge, which also causes torsion. In addition, the load evaluation allowed for oblique wind (at 45° and 60°), which had not been considered for the original design.
- A traffic restriction (no trucks) imposed when the third lane was implemented decreased the live loads carried by the trusses to approximately 57% of the original design load. This decrease largely offset the increase in dead load.

Results of the evaluation indicated that the main load-carrying members (main cables, hangers, towers, cable bents, stiffening trusses, floor beams, and stringers) have sufficient capacity (D/C ≤ 1.0) under all load combinations involving self-weight (dead load) and traffic (live load). The main cables have a maximum theoretical D/C of 1.02. This is not considered significant, and no actions were recommended.

D/C ratios are found to exceed 1.0 for the truss chords in the center span under the load combination with dead load, cable stretch, thermal effects, and wind load. This does not mean that the bridge is unable to carry its wind loads, but it does mean that the safety margin may be less than that contemplated by the current CHBDC. The bridge has withstood many storms safely. It is expected that where the D/C ratio exceeds 1.0 in the center span, the worst case would be localized damage. Since the bridge is closed during high-wind events, it is likely that no traffic would be on the bridge when localized damage may occur; therefore, it was decided that no action was required.

The deck lateral bracing was assumed to act as "tension-only" members and, in general, have D/C < 1.0. However, the end laterals in the panels adjacent to the towers and cable bents need to resist both tension and compression loads, and they are found to have D/C ratios of 1.29 at the ends of the main span and 1.57 at the cable bent end of the side spans. It was decided that this would be monitored, but that no reinforcing would be conducted because the bracing would be replaced along with the entire deck as part of the suspended spans' deck replacement in less than 10 years. Both cable bents have D/C ratios less than 1.0 except for the top struts, which will be strengthened in 2015 prior to the suspended spans' deck replacement.

8.6 MAINTENANCE

8.6.1 Annual painting program

Annual maintenance painting program has taken place on the Macdonald Bridge every year since the bridge was completed in 1955 except for two seasons in the early 1960s. As a result, the bridge steelwork remains in very good condition.

The painting has consisted of localized touch-ups only. The original paint system was an oil alkyd three-coat lead-based paint. The paint removal is done largely by using needle scalers. Due to environmental concerns with lead-based paints, in 1993 a zinc hydroxy phosphite three-coat oil alkyd system was introduced to replace the lead-based paint. The painting program is a seasonal operation carried out from May until early October each year and performed by a seasonal paint crew who has been mostly trained on the job (Figure 8.5). Due to maritime Canadian winters, maintenance painting is not practical to conduct during the winter. Bridge access for painting is by catwalks, temporary staging, permanent staging, wooden planks, bosun's chairs, and travelers. As safety systems and requirements have improved over the last 20 years, providing safe access has become more complicated to deliver.

Temporary horizontal lifelines are used extensively, along with permanent horizontal lifelines and anchor points. Full-body five-point harnesses and lanyards are worn by all painters at all times. All equipment must be tied off and, wherever possible, work is done from a solid floored system.

Paint repairs tend to have a shorter life than the original paint, and in many cases, it is the repaired paint that needs to be repaired. This is due in part to environmental conditions and the degree of surface preparation. While the bridge steelwork is in excellent condition, where the paint has been repaired there can be more than 10 coats of paint at the perimeter. Because of the widespread repairs and the excessive paint thickness, the usefulness of the painting program is becoming diminished. The painting program will evolve to complete removal and replacement in the near future and be implemented in phases to control cost.

In 1999, a third lane was added to the bridge (Figure 8.6) using an orthotropic steel plate deck on the approach spans. The new deck was painted with epoxy coatings. For the most part this approach was successful, but there are many areas where the paint has failed at sharp edges and the bond failed between the finish coat and the primer. This is being gradually addressed by removing the sharp edges and reapplying paint. It is noticeable that the shop-applied paint has performed much better than the field-applied paint.

8.6.2 Ice and snow control

Nine maintenance workers (four year-round employees and five seasonal winter employees) are primarily responsible for ice and snow control on

Figure 8.5 Painters at steel pier leg.

the Macdonald and MacKay Bridges. There are others who are trained and able to step in when there is a significant event.

The maintenance workers provide 24 hour-a-day on-site coverage from early November until late April. A small fleet of trucks equipped with salt–brine mixers and salt spreaders, along with rubber-tipped snowplow blades, provides the primary means of ice and snow control. Pay loaders or front-end loaders are used for clearing snow in the toll plazas and for moving snow to storage areas. A tractor with a salt spreader trailer is used on the sidewalks and bikeways of the Macdonald Bridge. The rubber-tipped blades are used to minimize possible damage to expansion joints and the road surface.

The bridges are also equipped with an ice-prediction system that includes a weather-monitoring station, road temperature and moisture sensors, and computer software that predicts when ice will form on the road deck. The

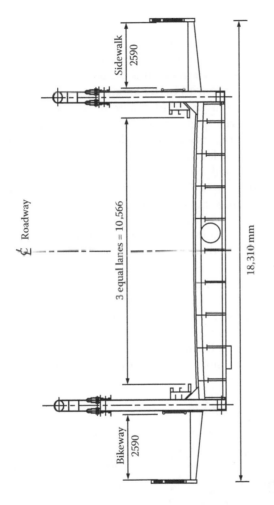

Figure 8.6 Section through suspended span after third lane project in 1999.

system allows for brine deicers to be placed before ice forms on the road deck.

8.6.3 Pier concrete repairs

Pier repairs were performed in the 1980s and in the 1990s and the third program has recently started. The original concrete placement used aggregate from riverbeds, non-air-entrained concrete, and smooth rebar. As a result, the concrete has suffered from AAR, which causes slow expansion of the concrete, significant map cracking, and accelerated deterioration of the concrete, along with significant strength loss.

Initial concrete pier–repair programs included localized removal and replacement of the concrete with better-quality concrete. In 2012 an extensive concrete investigation program whereby concrete cores were taken from each of the concrete piers, as well as the concrete abutments, was undertaken. This was done as part of the due diligence in the lead-up to the suspended spans' deck replacement.

The investigation found ongoing AAR and loss of concrete strength extending to a depth of 1 ft (300 mm). The testing of the cores revealed that the AAR has nearly exhausted itself. The recommended course of action was to remove the exterior face of the concrete piers and encapsulate the remaining concrete (Figure 8.7). Repairs were categorized into short, medium, and long-term and are being addressed over the next several years.

Concrete anchorage repairs took place in 2012 and a waterproof system was applied to the exterior face of the concrete. Water infiltration and high humidity in the anchorages have been an issue. Waterproofing of the anchorages was completed with the knowledge that dehumidification of the main cables and anchorages would take place as part of the upcoming suspended spans' deck replacement. The waterproof material has cracked in several locations and moisture within the anchorage remains an issue.

(a)   (b)

Figure 8.7 Dartmouth cable bent concrete pier (a) before and (b) after repair.

The cracking of the waterproof is being investigated as part of the warranty. The manufacturer was not surprised that the waterproof cracked, in spite of the waterproof being reinforced on all faces. It is understood that the plan will include application of additional waterproof and reinforcing where the waterproof has cracked. Waterproof has been used on other concrete piers on the MacKay Bridge with poor results. Thickness would seem to be a critical aspect of waterproof application. It is too early to decide if this latest attempt of waterproofing concrete will be successful.

In 2014, while repairing an approach pier on top of the Halifax anchorage, a horizontal crack was found just above the anchorage roof. It was determined that the crack is active. It appears that the bearings for the approach girders are seized. A separate program was already underway to assess the plate girder bearings, as it was suspected that they were seized. The bearings are from the original construction of the bridge. For this pier, temporary crack control and support was designed and added as part of the concrete repairs. The seized bearings will be replaced shortly.

8.6.4 Wearing surfaces

The Macdonald Bridge originally had a concrete deck on the approach spans and a concrete-infilled grid deck on the suspended spans. By 1970, there was significant deterioration of the concrete surface on the approach spans and wearing of the concrete between the steel grid on the suspended spans. After the MacKay Bridge opened in 1970, a concrete-repair program was carried out on the Macdonald Bridge in 1971. After concrete repairs, a thin layer of epoxy asphalt was applied. The epoxy asphalt did not perform well and was subsequently replaced in 1976 with a thin layer of asbestos-modified asphalt.

When the bridge was widened from two lanes to three lanes in 1999 and the approach spans deck replaced with orthotropic steel plate deck, the asbestos-modified asphalt was removed. The asbestos-modified asphalt performed well, even though it had worn significantly in the wheel tracks. A nominal 0.375 in (9.5 mm) thick epoxy urethane wearing surface was installed on the new steel deck on the approach spans and on the grid deck on the suspended spans. The thin wearing surface was used to minimize the weight on the suspended spans. While the epoxy urethane wearing surface was advertised as waterproof, it proved not to be, with significant areas losing bond to the steel deck and failing, and wearing/polishing rapidly in the wheel tracks resulted in reduced traction.

In 1999, long-term samples of five other products were placed on the steel deck and on the concrete-infilled grid deck. One product showed very good performance and was selected to replace the failing epoxy urethane wearing surface. The wearing surface on the approach spans was replaced in 2005 with a nominal 0.375 in (9.5 mm) thick epoxy wearing surface. On the suspended spans the existing wearing surface was roughened and the epoxy wearing surface was applied on top of the existing epoxy urethane. The original 1999 sample of the epoxy wearing surface was left in place. The epoxy wearing surface showed premature surface wear and loss of traction; however, the 1999 sample continued to perform well.

Ultimately, it was decided in 2008 to replace the thin wearing surfaces on the approach spans with 2 in (50 mm) of polymer-modified asphalt on top of a waterproof system. The epoxy wearing surface was left in place on the suspended spans. Traction continued to be an issue and so the surface in the wheel tracks was grooved using diamond blades as a trial. The grooving lasted for about 2 years but ultimately it too resulted in poor traction. As a short-term solution, an asphaltic microsurfacing was placed. The surfacing will be completely replaced as part of the suspended spans' deck replacement.

While thin wearing surfaces are tempting because of weight savings, they cannot be relied upon to be waterproof and are particularly susceptible to premature loss of traction. It is critical with thin wearing surfaces that there be a neat application of the material as the base layer to improve the chances of the surface remaining waterproof. The cost of wearing surface failures is not just the capital cost of replacing them but also the disruption to traffic and loss of public image.

8.7 REHABILITATION

The Macdonald Bridge has undergone routine repairs to concrete road decks, wearing surfaces, and concrete piers. The most significant rehabilitation was the replacement of the approach spans concrete deck with orthotropic steel plate deck in 1999. At the same time the deck was widened from two to three lanes (Figures 8.8 and 8.9) and all of the expansion joints were replaced.

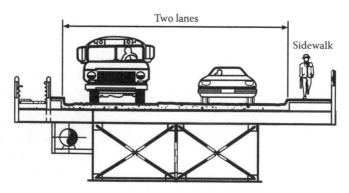

Figure 8.8 Old deck system on approach spans.

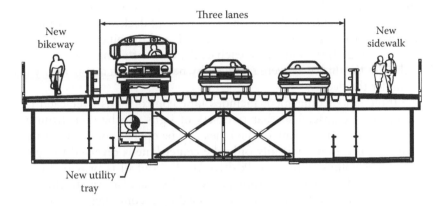

Figure 8.9 New deck system on approach spans.

8.7.1 Third lane project

In the late 1980s and early 1990s, an annual program of repairing spalls in the top surface of the concrete deck of the approach spans was required. Underside repairs were conducted in a few locations and steel retaining plates were added underneath as well. The area of repairs was increasing and it was apparent that the deck required replacement.

The Macdonald Bridge was quite congested with traffic and there was insufficient capacity on alternate routes to allow the bridge to be closed for an extended period. This required that the deck replacement be conducted during off-peak periods using prefabricated deck panels. Orthotropic steel plate deck was selected since it could immediately be put into service.

It was desirable to increase cross-harbor traffic capacity. The feasibility of adding a third lane to the bridge was examined and confirmed to proceed. At the same time, a bike path was to be added.

To achieve three lanes, it was necessary to ban trucks and minimize the added dead loads to the bridge. Because the bridge connects to city streets, the times during which trucks could use the bridge was already limited, so a complete ban on trucks was readily achieved. There was some synergy between limiting additional dead loads and providing prefabricated deck segments since the orthotropic steel plate deck was lighter than traditional cast-in-place concrete. However, the need to minimize dead load effectively eliminated precast concrete panels as a possibility.

Three lanes were achieved on the approach spans by providing a wider deck, with sidewalk and bikeway panels that were bolted on after the road deck was installed, for ease of handling. The utility cables in the duct way were temporarily supported and ultimately replaced with new cables in a new utility tray slung on the underside of the deck (Figure 8.9). It was required that the copper core cables be replaced with lighter smaller fiber-optic cables to reduce dead load.

On the suspended spans, the sidewalk and duct way were removed and the road deck was extended using the same type of deck as the existing steel grid deck infilled with concrete. The sidewalk and the bikeway were added by cantilevering off the outside of each of the two stiffening trusses. Again, the cables in the duct way were replaced by new cables in a new utility tray slung under the deck.

The three-lane arrangement includes lane control signs over each lane. The gantries are spaced so that two sets of signs can be seen from anywhere along the bridge. Any of the lanes can be opened, or closed, to traffic in either direction. Typically only the center lane is reversed to operate in the higher-volume direction. Positive traffic control is provided on the approaches to the bridge by closing gates to block access lane to the center lane. The system can be centrally controlled by a computer system in the operations center at the bridge.

The approach spans' deck replacement and the addition of the third lane was constructed from 1997 to 1999. Prior to construction, the wait time to cross the bridge during peak periods was on the order of 40 min. Following installation of the third lane, the wait time was reduced to only a couple of minutes, due more to signalized intersections on either end of the bridge.

8.7.2 Tieback system

A survey and analysis of the bridge was carried out in 2006. Some anomalies were noted in the position of the towers and cable bents and the deck profile. There was a concern that the factor of safety against slippage may be less than desirable between the main cable and the top of the cable bent. The factor of safety was improved by introducing steel cable tiebacks attached to the top of the cable bents and routing them to an anchoring system in the anchorages. In this way, the top of the cable bents could not shift toward the midspan.

8.8 BEST PRACTICES

8.8.1 Security system

HHB implemented a security system that includes over 100 intelligent cameras at strategic locations on the bridges and surrounding lands.

The system is primarily for security, but it is also used for monitoring traffic. The operators have the ability to change the position of cameras and zoom in and out, but the cameras return to default views after a time-out period.

The system includes analytics on the images, establishment of restricted zones or perimeters, and setting of alarms, so that the operator does not have to look at the screens continuously. When an alarm is issued, the appropriate camera display is brought to the front and enlarged on the monitors, along with text/flashing warnings and an audible alarm to draw the attention of the operator. Other cameras in proximity of the alarm's location change position to focus on the area from other vantage points and these images are displayed as secondary images. This advanced system enables the operations team to process incidents quickly and assign resources efficiently.

8.8.2 Maintenance and construction disruptions

HHB schedules maintenance and capital works with the goal of minimizing traffic disruption. HHB uses as a general rule that traffic disruptions are a last resort. Work is performed from staging or land-based access when possible to avoid disruption to traffic, including road, pedestrians, cyclists, and marine craft. When disruption is absolutely necessary, it is scheduled for off-peak periods when the impact is lessened. HHB coordinates multiple projects to take full advantage of any disruption.

In general, all roadway lanes remain fully open between 5:30 AM and 7:00 PM. At least one roadway lane is open each way weekdays from 7:00 PM to 5:30 AM and on weekends from Friday at 7:00 PM to Monday morning at 5:30. Lane closures are not to be permitted on the Macdonald Bridge when the MacKay Bridge is closed or has lane reductions or vice versa. Sidewalks and bikeways can be narrowed during off-peak periods. Generally, when sidewalk or bikeway is closed, the other remains open. However, when both facilities are closed, a shuttle service is provided. Work activity that conflicts with marine traffic requires coordination with marine traffic controllers in the harbor and the access is made mobile so that it can be moved out of the way to allow vessels to pass.

It is HHB's policy to begin temporary works' traffic control in the plazas at the respective ends of the bridge, regardless of where the work is taking place between the plazas. This reduces the number of merges to fewer lanes that otherwise would occur on the bridge, where there is no escape lanes or shoulder.

8.9 OTHER RELEVANT ITEMS

8.9.1 Highlighting

On January 2, 2000, passive white highlighting was turned on to celebrate the millennium and the 45th birthday of the Macdonald Bridge. The highlighting consists of floodlighting of the towers at the pier and deck levels. Initially the highlighting included floodlights along the top of the stiffening trusses to light up the main cable; however, this was found to be largely ineffective because of the orange color of the cable and the slender surface.

To improve the effectiveness of the highlighting on the main towers, the dark-green paint was changed to a lighter green with a higher sheen in 1999.

HHB selected the passive highlighting to focus attention on the most attractive feature of the bridge, the towers. The lighting was selected through computer modeling, renderings, and a field trial.

8.9.2 Reversible lane

With the introduction of the third lane in 1999, a means of reversing the center lane was required. A series of overhead signs indicating X and arrows control whether the lane is open or closed for that direction of traffic. In concert with these overhead signs are additional variable message signs at the abutments of the bridge and gates to physically constrain the traffic to one lane or to open the center lane. The lane is typically changed at midnight and noon. The lane is routinely reversed for special events and as dictated by traffic conditions, where the security cameras are used as an aid.

8.9.3 Air gap–measuring system

Halifax has one of the deepest ice-free harbors in North America and is the reason that the city was established. It is important to ensure that the working harbor remains functional. As vessels have increased in size, the clearance for navigation under the bridge has become a concern. As a joint initiative with the Halifax Port Authority, HHB established an automatic identification system (AIS) beacon on the Macdonald Bridge that with the aid of GPS, tide gauges, and tide charts provides the available vertical clearance under the bridge to pilots on the vessels moving in and out of the harbor. AIS beacons are an automatic tracking system used on ships and by vessel traffic services for identifying and locating vessels. It is used to supplement radar.

The provision of AIS enhances navigation, reduces the risk of a collision and allows vessels to travel under the bridge confidently with as little as 4 ft (1.2 m) of vertical clearance. To further aid navigation, the bridge has a whiter roadway light at midspan and orange panels at 180 ft (55 m) offsets from midspan. These visuals are to assist the pilots in navigating the channel beneath the bridge and for them to center the highest part of the vessel on to maximize the vertical clearance.

8.9.4 Tolling systems

The operation and maintenance of the bridge is funded exclusively from tolls to cross the bridge. On the Macdonald Bridge the tolls are collected in 10 lanes (5 in each direction), and a toll is charged for each trip at $1.00 cash for passenger vehicles or $0.80 by electronic tag. In 1955, all lanes were serviced by a toll operator, who collected payment by ticket or cash and later by token or cash. Since the introduction of electronic tolling in 1998 (MACPASS) the number of MACPASS-only lanes has increased to two in each direction and the number of service lanes has been reduced to one in each direction from two. MACPASS is accepted in all lanes.

The tolling system uses a passive tag, requiring no battery in the tag. Over 73% of all trips are made using electronic tolling. The lanes are equipped with gates and patron fare indicators to provide feedback to the driver and to minimize nonpayment.

A study is currently underway to assess if and when an all-electronic tolling system should be implemented.

8.9.5 Weigh-in-motion scales

Weigh-in-motion scales were installed in the early 1990s in the MacKay Bridge toll plaza to weigh heavy trucks. Because of the stop-and-go conditions in the toll plaza, it was found that the weigh scales did not perform well and the results were questionable. As a result, the weigh-in-motion scales were abandoned and eventually removed.

8.10 FUTURE PLANS

8.10.1 Suspended spans' deck replacement

The Macdonald Bridge is about to undergo a major refurbishment with the complete replacement of the suspended spans deck, including the deck, stiffening trusses, floor beams, the sidewalk, the bikeway, the water pipe, and hangers. Due to corrosion between the deck and the supporting stringers, along with deteriorating ride quality as a result of corrosion, the deck must be replaced. The purpose of the project is to extend the life of the bridge and reduce the dead load, which was increased in 1999 with the addition of the third lane. One of the criteria for the project is to provide a lower-maintenance structure and enhance safe access.

Following extensive study, analysis, and design, a contract was awarded for the replacement of the suspended spans deck in 2014. The suspended structure will be replaced in 66 ft (20 m) segments lifted from a barge in the harbor below and 33 ft (10 m) segments lifted from on the bridge where over land. The work will be done substantially at night to permit the bridge to be open to traffic during the workday.

The new suspended structure will be the same width as the existing structure, with three roadway lanes, a sidewalk, and a bikeway. The trusses will be positioned below the deck rather than above it (Figure 8.10), offering greater protection from salt spray from passing vehicles. The new suspended structure will eliminate the gap in the plane of the existing

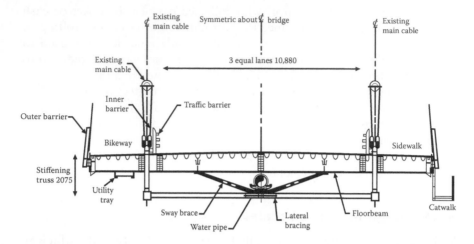

Figure 8.10 Section through suspended span redecking project planned for 2015 construction.

stiffening trusses between the roadway and the sidewalk/bikeway. The new stiffening trusses will be made of hollow structural members that will be sealed, rather than built-up, members as on the existing stiffening trusses, to minimize the surface area requiring paint and maintenance. The deck will be an orthotropic steel plate deck.

Prior to replacement of the deck segments, the sidewalk and the bikeway must be removed and the 24 in (600 mm) water pipe must be emptied and removed in select locations. Existing stiffening trusses must be reinforced because the loads introduced with the temporary deck connection between the new and old stiffening trusses. The traveler rails will be reinforced so that the temporary deck connection can be rolled along to the next segment during the overnight segment replacement. The new segments, when delivered, will include barriers, sidewalk, bikeway, water pipe, and prepaving, but will still weigh less than the existing three-lane roadway segment. Therefore, concrete ballast will be provided with the new deck segments to maintain the profile of the bridge during construction.

To improve the ability of contractors to bid the project, reduce the construction schedule, and ensure the safety of the public, the erection sequence and major erection equipment were designed by the owner's engineer prior to issuing the tender. Because the owner's engineers had to carry out detailed investigation, analysis, and modeling, they were in the best position to design the erection sequence and major erection equipment.

The segment replacement process includes positioning of a lifting gantry above the roadway deck, which will be hung from the existing hangers and, in turn, supports the existing deck during removal. The roadway deck is cut at the limit of segment being replaced, and the lifting gantry takes up the weight of the segment and adjusts so that there is zero force in the chords of the stiffening trusses and, therefore, they may be cut as well. The existing segment will be lowered and a new segment will be raised from a barge in the harbor. The new segment will be bolted to the previously installed new deck segment to allow the bridge to be reopened to traffic. On a subsequent night the transverse deck splice will be welded between the new deck segments.

To connect the new deck segment to the existing deck, a temporary deck connection is required between the new and existing stiffening trusses. When the new deck segment arrives on site, a nosing will be bolted to the front. On the existing deck segment the temporary deck connection, mounted on the traveler rails, will be rolled back to the next segment location to be replaced and connected to the existing stiffening truss, and when the new deck segment is in position, the two parts of the temporary deck connection with be connected.

Fabrication is underway for the deck segments, with installation planned to begin in August 2015. Replacement of the deck is anticipated to take about 13 months. Several other small projects will be completed as part of the suspended spans' deck-replacement project and they are described in the following.

8.10.2 Hand ropes on main cable

At the beginning of the suspended spans' deck replacement, the contractor will install hand ropes on the main cables for access to the cable bands at each of the hangers, to check cable band bolt tensions, and for replacement of cable bands in select locations. In the long term, HHB will use the hand ropes for inspection and maintenance access.

Currently a cable crawler is used on the main cable for access, painting, and maintenance. The cable crawler rides on the main cable and pulls itself up via a cable anchored at the top of the tower. The cable crawler will cause excessive stress in the cable wrap system to be installed with the dehumidification system and may damage it; therefore, the cable crawler will no longer be used.

8.10.3 Dehumidification

After the suspended spans' deck replacement, the main cable will be wrapped with a long-life elastomeric wrap, and dehumidified air will be pumped into the cables at midspan (Figure 8.11). Exhaust points will be at the anchorages where the flow and humidity will be monitored. The goal is to reduce the humidity in the cable to below 40% to effectively arrest the possibility of corrosion and indefinitely extend the life of the main cables.

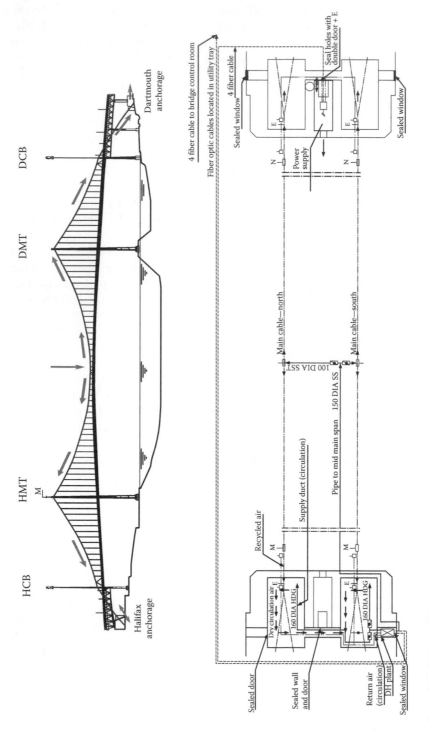

Figure 8.11 Dehumidification system schematic.

8.10.4 Safety systems in conjunction with travelers and access systems

When the bridge was built in 1955, safety was not as much of a concern for maintenance workers. At that time it was expected that there would be a number of deaths on a large project. The approach and attitude about safety has changed from expecting deaths to implementing plans and procedures with the intent of there being no injuries. A significant focus of the design of the suspended spans' deck replacement is to engineer out safety issues.

Openings in barriers have been specifically designed to ensure that small children cannot slip through and to eliminate handholds or footholds on the upper part of outer barriers that might assist in climbing over the barriers.

For the maintenance of the bridge, access platforms and catwalks have been provided below deck at the towers and cable bents. A headroom, clearances, and emergency access have been considered. Across the suspended spans a traveler will run on rails below the deck, but for emergency evacuation in the case of a traveler breakdown, a narrow catwalk has also been provided. The initial emergency access plan to the travelers was to climb over the outer barriers, but the outer barrier design is so effective that, even with ladders, it would be impractical.

Anchor points and bosun's chair rails will be provided on the towers and cable bents. Access hatches and ladders have been designed for stretchers to be lifted through them, including specific anchor points for pulleys. Hatches have been designed so that they will not fall on a person climbing through, plus guards will be in place at the top of hatches to protect someone to walk through an open patch. Wherever possible a door is used rather than a hatch.

Equipment platforms and access platforms at the towers are screened in with netting to prevent birds from roosting in the area. Birds and bird feces have proved problematic on the existing bridge.

8.10.5 Light-emitting diode lights

Light-emitting diode (LED) roadway lights will be used on the suspended spans, as well as the approach spans, to improve energy efficiency and minimize light pollution.

The existing highlighting will be replaced with new fixtures, which will also be LED.

To further enhance marine navigation of the bridge, sector lights and beacons will be added to the underside of the bridge near midspan. The existing aerial navigation beacons will be replaced at the same time.

Where possible, LED fixtures will be used for reduced energy consumption, light pollution, and maintenance.

8.10.6 Replacement of approach spans' bearings

One set of bearings are seized on the approach spans' plate girder sections. Prior to discovery of the seized bearings, a monitoring program had been initiated and design is underway for replacement of the approach plate girder span bearings. The approach truss bearings are functioning adequately. The bearings will be replaced over the next couple of years by jacking up the deck on temporary bearings, retrofitting the bearing seat, and installing the replacement bearings.

8.11 CONCLUDING REMARKS

The Macdonald Bridge is a key crossing of the Halifax Harbour, providing an essential connection in the Halifax transportation network. The 60-year-old structure, with approximately 45,000 crossings on an average workday, is well utilized.

As the organization responsible for the operation and management of two long-span bridges across Halifax Harbour, the mission of HHB is to provide safe, efficient, and reliable cross harbor transportation infrastructure at an appropriate cost.

As the stewards of the bridges, the HHB team takes pride in ensuring that the bridges are well maintained year round and disruptions are mitigated by accommodating commuter traffic, and long-range planning ensures a long useful life of the bridges for future generations.

REFERENCE

1. Chapman, H., 2005, *Crossings: Fifty Years of the Angus L. Macdonald Bridge*, Nimbus Publishing Limited, Halifax, Nova Scotia, Canada.

Chapter 9

Mid-Hudson Bridge

William J. Moreau

CONTENTS

9.1 OPENING DAY

Eleanor Roosevelt cut the ribbon to open the Mid-Hudson Bridge (MHB), on August 30, 1930 (see Figures 9.1 and 9.2). The 1500 ft main-span suspension bridge had been sought after by local businesspersons and community leaders for many years (see Table 9.1). The economy of the Hudson Valley relied on the transportation resource provided by the Hudson River, but times were changing and vehicular transport in the United States was about to explode.

9.2 DESIGN AND CONSTRUCTION

Designed by Modjeski and Moran, the bridge was constructed by American Bridge. While a huge majority of suspension bridges, built prior to the MHB, had used high-strength bridge wire manufactured by the Roeblings, this bridge would be built with wire manufactured at the new American Bridge wire factory in Bethlehem, Pennsylvania (see Figure 9.3).

Figure 9.1 MHB, Poughkeepsie, New York.

Figure 9.2 Ribbon cutting for the MHB, August 30, 1930.

The location for the bridge was selected by Ralph Modjeski. The Poughkeepsie landing was the subject of some debate. Modjeski would have preferred the bedrock bluff, just south of the current bridge location (Kaal Rock), but he agreed with the locals to save a prestigious inn located on the Kaal Rock bluff by relocating the east abutment to the adjacent

Table 9.1 MHB fact sheet

Type of bridge	Suspension
Construction started	May 15, 1925
Opened to traffic	August 25, 1930
Length of main span	1495 ft
Length of side spans	750 and 755 ft
Length, anchorage to anchorage	3000 ft
Total length of bridge and approaches	9380 ft
Number of traffic lanes	3 (reversible center lane)
Width of roadway	31 ft
Height of towers above mean high water	315 ft
Clearance at center above mean high water	135 ft
Depth of caisson foundation	135 ft (east), 115 ft (west)
Number of cables	2
Diameter of each cable	16 3/4 in
Diameter of suspender ropes	1 3/4 in
Total number of wires in each cable	6080 galvanized no. 6
Structural material	Steel, carbon, and silicon
Tower material	Steel and granite
Deck material	Concrete-filled steel grid (1987–1988)
Cost of original structure	$5,890,000

Figure 9.3 Main cable spinning over the completed towers of MHB.

bluff, 500 yd to the north; see Figure 9.4. Construction commenced with the deep-water foundations for the river piers through 50 ft of river water and 70 ft of soft silt to reach high-capacity compacted gravel for support of the tower foundations.

Open caissons were used to protect the deep excavations from the river. A two-dredge system was developed to speed the construction; one clamshell

Figure 9.4 Tower erection in progress; eastern bank of the river in background.

excavation bucket was mounted to a crane located on barges at each end of the 50 ft × 100 ft caisson. Work progressed simultaneously until one of the cranes excavating for the east tower broke down. Not wanting to further delay work progress, the second crane continued working through the day.

When the crew arrived at the job site the next morning, they were met with a devastating sight: the east river caisson was listing at a 25°–30° angle. Work could not continue as the clamshell bucket would no longer drop vertically through the many cells forming the caisson. Months would pass and winter set in before a plan was developed to right the leaning caisson. When the foundation was finally corrected, it was found to be 5 ft west of the intended position. Modjeski never acknowledged this publically; even the shop drawings went unchanged and modifications to the fabricated steel were made in the field. The actual main-span distance is 1495 ft, the east side span is 755 ft, and the west side span is 750 ft. Virtually all historic records of the MHB identify its main span as 1500 ft, making it the world's sixth longest suspension bridge when it opened.

9.3 NEW YORK STATE BRIDGE AUTHORITY

The bridge was built by the Department of Public Works for the state of New York, with the original authorization made by then Governor Franklin D. Roosevelt. Roosevelt became a U.S. president before the bridge opened under Governor Smith, who invited the first family to return to cut the ribbon. Two years later the New York State Bridge Authority (NYSBA) was created to build the Rip Van Winkle Bridge across the Hudson River at Catskill, New York (see Figure 9.5). The state gave MHB to the bridge authority, which then committed the future revenues of both bridges to the private Wall Street investors through a public bond offering. This method

Figure 9.5 The Rip Van Winkle Bridge (Catskill, New York) opened in 1935; the total length is 5000 ft.

of funding public infrastructure through private financing led the way for many future quality-of-life improvements, including, transportation, potable water supply, flood control, and energy-related projects.

9.4 MAINTENANCE AND INSPECTION

Maintenance and operations were the primary responsibilities of the bridge authority. Major repair or bridge reconstruction remained primarily with the state as the early public authorities were not well staffed with experienced engineers. As the suspension bridge industry matured and expanded, the expertise of building and maintaining long-span suspension bridges remained with only a few individuals. Only one major suspension bridge opened in the United States during the second half of the 20th century, and this was the Verrazano–Narrows Bridge in New York City, circa 1964. The next major suspension bridge built in the United States was the Carquinez Bridge in Vallejo, California, in 2003. This lack of opportunity limited the exposure and experience of bridge engineers to the nuances of suspension bridge construction, inspection, maintenance, and repair. The NYSBA sought to address this issue by hiring the firm of Ralph Modjeski to perform annual inspections of their bridges to address ongoing maintenance items and to plan for major investments in component replacement or rehabilitation. These regular inspections date back to the late 1950s upon completion of the Kingston–Rhinecliff Bridge in 1957. Although the National Bridge Inspection Standards [1] had not yet been developed at that time, the NYSBA recognized the value and importance of planning for regular maintenance, repair, and timely reconstruction of these valuable public investments.

The first main cable inspection of the MHB, beyond the visual inspection accomplished by simply walking the cable, was made in 1974 (see Figure 9.6). This means that the main cables were already 45 years old (1929–1974)! Our experience now indicates that this is too late in the life of a main suspension bridge cable to begin logging and addressing environmental, chemical, and molecular wire degradation.

The MHB was wrapped with three plies of soft number 9 galvanized wire. In this case, three plies means three pieces of wire were wrapped circumferentially around the main cable at the same time, not three layers deep as a ply would infer in automobile tire manufacture. This meant one ply of wrapping wire could be cut and removed, leaving the other two plies in place, maintaining the original level of cable compaction.

This 1974 inspection revealed that the main cable wires were in generally good condition, with modest amounts of white corrosion product (oxidized galvanizing), dried red lead paste, and virtually no ferrous corrosion. No broken wires were detected.

Five years later, in 1979, a 6 in length of wrapping wire was removed, all three plies, so that a more detailed look at the parallel wires could be made. During this inspection, significant ferrous corrosion was detected for the first time (50 years), indicating complete loss of the galvanic corrosion protection layers in localized areas. The higher level of corrosion in some of these areas made sense (e.g., the 6 o'clock position of the cable cross section). Other areas of intensified corrosion were not so easily understood. Black spots and corrosion pits were noted on random wires during the cable inspections in the early 1980s; see Figure 9.7. Each inspection led to

Figure 9.6 Two plies of a three-ply wire wrapping were removed for inspection, MHB, circa 1974.

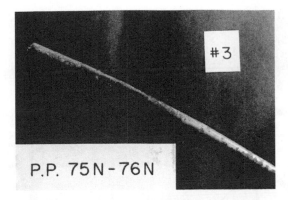

Figure 9.7 Corrosion pit and significant section loss of wire sample.

a recommendation to complete additional invasive cable investigations in order to collect a larger pool of data, sampling both the good and deteriorated wires for clues as to how to better protect the main cable system.

9.5 MAIN CABLE EVALUATION

Since there was no established testing or evaluation procedures for deteriorated suspension bridge cables, the NYSBA retained two highly respected suspension bridge design consultants to evaluate the MHB and the Bear Mountain Bridge independently and then compare and contrast results and procedures to establish a level of confidence in predicting the current safety and future useful life of these critical bridge components. Upon completion of their in-depth cable inspection and wire sampling and analysis, neither consultant was comfortable predicting the main cable capacity beyond a 5-year window. With such limited experience in this area, the authority agreed with their consultants and the authority began invasive main cable inspections on a 5-year cycle beginning in 1990.

Extensive material testing was performed on main cable wire samples in various investigations at both authority suspension bridges during the 1980s; see Figure 9.8. While fatigue testing failed to uncover any new understanding of fracture mechanics, the pull tests and stress–strain results indicated that most, if not all of the cracked wires, had elongation results well below the minimum specified for the wire during original construction. Why was the wire becoming brittle? Why some and not all wire samples? Was there a correlation between brittle wires and the extent of corrosion? The answer to the last question was easily found. It was no. Many highly corroded wires maintained their ductility during testing. Despite the lack of correlation, it was clear that a highly corroded wire has a reduced cross section, yielding a higher working stress and leading to a higher potential

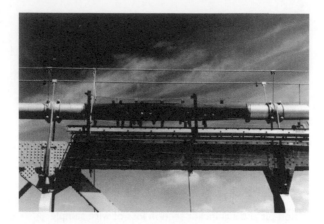

Figure 9.8 Main cable wedging of MHB, circa 1980.

for crack development and growth. Both consultants designed computer models to apply statistics to the unknown quantity of main cable wires based upon the distribution of deteriorated wires detected during the invasive cable inspections.

The open dialogue between consultants and bridge owners proved fruitful; both consultants' methods yielded similar results and held similar assumptions in their calculations. While this gave some degree of confidence to the current calculated level of safety, no one was confident in predicting the future. An all-out effort began to improve the level of corrosion protection for the main cables.

9.6 MAIN CABLE CORROSION PROTECTION

The in-depth cable inspection did note that individual wires coated with a dried linseed oil film appeared to be in better condition than wires without any linseed oil deposits. Advice was sought from other corrosion prevention experts, from the mining industry to antenna tower operators.

The consultant for the MHB requested the opportunity to wedge open and inspect 20% of the length of each main cable before they would render a professional opinion on the continued safe use and future useful life of the bridge. This was the case due to the lack of correlation between cracked and embrittled wires with any other observable factor. Accordingly, we, as bridge owners/operators, wanted to be proactive while this rare opportunity existed to get inside the main cable. Various corrosion inhibitors were applied and evaluated, including vapor phase inhibitors, single- and double-boiled linseed oils, and petrolatum; see Figure 9.9.

After some time and trials, a new product was offered by the corrosion protection manufacturer combining all desired properties and performance.

Figure 9.9 Application of corrosion inhibitors to the main cable of the MHB.

A double-boiled linseed oil with chemical dryers and thinned with turpentine to the optimum viscosity was selected for application to all exposed wires. It was allowed to saturate the cable by filling in the open wedge lines for 12–24 hours.

Upon completion of the inspection, oiling, and rewrapping of the 20% length of cable, it was determined that the best move forward was to open and oil the balance of the cable length. This work was completed in 1994. No active corrosion in the main cables has been observed during the three inspections (15-year duration) since the oiling was performed.

The NCHRP Report 534 [2] manual was published near the end of the 20th century and has been used to evaluate the cable strength and safety in the most recent cable inspection. The resulting safety factor is slightly better using the NCHRP method [2] as compared with the independent consultant-derived models. The BTC method [3] has also been used to evaluate the Mid-Hudson. All methods have yielded consistent results, within approximately 15% of each other.

Red lead paste was used below the wrapping wire during the original cable construction and the work of the 1980s and 1990s. During cable rehabilitation of the Bear Mountain Bridge, an improved paste made with zinc dust and zinc oxide paste was developed to replace the red lead; see Figure 9.10. All cable rehabilitation work in the United States now utilizes this zinc paste to improve the safety of the bridge cables and the bridge workers. The subject of this zinc paste and its development are discussed in detail in companion volume [4], in the "Main Cable Corrosion" chapter.

Next, the outer layer of protection beyond the wrapping wire was traditionally just paint. Various elastic paints were tested and these aptly outperformed the original alkyd oil–based paints. Hypalon, urethane, and waterborne acrylics performed the best. Partially cured ethylene propylene

Figure 9.10 Zinc paste substitute for traditional red lead paste.

diene monomer membranes and other tapes were tested as an outer cable wrap; see Figure 9.11. Sealing these preformed tapes can be challenging at each cable band. All systems evaluated outperformed the level of protection provided by alkyd paint. The MHB is currently painted with Noxyde, a waterborne acrylic paint; see Figure 9.12. This paint system has provided good protection for nearly 25 years and continues to perform well during this period of constant invasive inspections.

Suspender ropes were also evaluated during the main cable inspection to take advantage of the specialty support contractors on site to open and close the main cable. Three representative suspenders are selected for replacement and destructive laboratory testing. One long, one medium, and one short suspender ropes are replaced during each 5-year cable inspection cycle. After 80 years of service, some ropes are just beginning to show pull test results below the original design and construction specifications.

Figure 9.11 Neoprene cable wrap system by D.S. Brown.

Figure 9.12 Waterborne acrylic paint—Noxyde.

9.7 SUSPENDER ROPES (HANGERS)

The most common detrimental circumstance for suspender ropes is the connection to the stiffening truss or girder. The connections at the MHB are made at the top chord of a pony-style stiffening truss with the traffic near the bottom chord and no upper lateral bracing for the top chord; see Figure 9.13. This results in a relatively clean location away from traffic and unconfined for easy cleaning and painting. Bending of the end of the rope is softened at the top chord connection by the use of a cast-steel guide or cradle. The deterioration observed in the ropes, pull tested to failure, is typically located at the inner core wires. These wires are generally slightly smaller in diameter and can be exposed to damp conditions more than the outer wires.

Nondestructive testing equipment has been demonstrated in evaluating various suspender rope installations. While ultrasonic and magnetic flux technologies are effective at detection along the "run" of a rope, they are all ineffective

Figure 9.13 Suspender rope connection to stiffening truss at MHB.

at the cable band and truss connection. The NYSBA continues to evaluate all suspender ropes over 70 years old on a 5-year cycle, testing to destruction a representative sample of ropes, e.g., 3 out of 300. The NYSBA also does not use a heavy-bodied paint near the lower 5 ft of the rope. This allows any trapped moisture to escape before the socket. The upper lengths of the rope may be caulked, painted, and sealed, especially as they pass over the cable bands.

The only cables that were completely replaced on the MHB were the two safety lines mounted on either side of each main cable. These 1 in diameter lines were replaced at approximately 75 years of service. Improvements to the safety line posts at each cable band were also made along with the installation of a continuous line over the tower tops. The original lines were not carried across the top of each tower, complicating a thorough main cable inspection.

Cable band bolt tensioning was performed after the main cable rehabilitation work was completed and rechecked at 3-year intervals. If more than 15% of the bolts tested were in need of retensioning, then all bolts are scheduled to be tightened on a more systematic basis. A full description of the various types of cable band bolt tensioning equipment is discussed in companion volume [4]. Cable band bolt tensioning should be considered after cable wedging (performed during inspections) or on a 25-year maximum interval. The cable band bolts on the MHB were all replaced at approximately 65 years of service to facilitate the retensioning procedure.

9.8 ANCHORAGES

The anchorages of the MHB are quite different from one shore to the other. The eastern shore has a large masonry structure built on a bedrock bluff, reasonably well drained. The west anchorage is carved into the midheight wall of a large bedrock escarpment that runs parallel to the river; see Figure 9.14. Fissures in the bedrock feed groundwater directly into the west anchorage chambers.

Figure 9.14 Original construction of west MHB anchorage, circa 1929.

Figure 9.15 Typical desiccant wheel dehumidification equipment.

Fortunately, the upper chamber can be separated from the eyebar chamber and drain holes were cored to allow the trapped water to escape the upper chambers. The eyebar chambers of both anchorages were then isolated and outfitted with desiccant wheel–type dehumidifiers in the late 1990s; see Figure 9.15.

9.9 TOWERS

The tower tops were then dehumidified in 2002. The ornamental "hoods" over the main cable saddles of the MHB were found to be contributing significant amounts of moisture to the main cables. These hoods were originally well vented and their shape directed any condensation that formed on the interior surface to drip right onto the cable saddle. Ventilation was eliminated and higher security was added during the tower hood dehumidification project; see Figure 9.16.

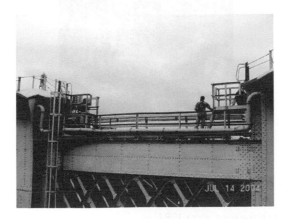

Figure 9.16 Dehumidification of the tower saddle hoods at the MHB.

It is important to control dehumidification equipment on a time-based cycle. While using humidity levels to control run time may seem to be the most efficient and effective strategy, long downtime will reduce the reliability of the hardware. Equipment should be cycled on at least once a day.

9.10 DECK REPLACEMENT AND GENERAL SUPPORTING INFRASTRUCTURE

Maintenance and rehabilitation of suspension bridges is not limited to the suspension system. The last major projects funded and managed by the state of New York's Department of Transportation at the MHB were the widening of the east approach roadway and the realignment and widening of the west approach roadway in the mid 1960s. The west approach roadway now follows a very deep rock cut through the escarpment forming the Hudson River valley. Geology professors and their students often visit this area for a unique cross-sectional view of the earth's crust.

All capital improvements are currently funded through the bridge authority toll revenue and the issuance of municipal bonds. The following is a list of the major rehabilitation projects that have been necessary to maintain the MHB in a state of good repair:

- 1987–1988: full deck replacement (see Figure 9.17)
- 1995: sidewalk replacement and traveler installation (see Figures 9.18 through 9.20)

Figure 9.17 Deck replacement, circa 1987.

Figure 9.18 Original sidewalk being removed (steel bracket widened for new fascia stringer).

Figure 9.19 New sidewalk, circa 1995.

- 1996–2008: lead removal and full repainting of steelwork (see Figure 9.21)
- 2009: safety line replacement along the main cable
- 2012–2014: tower wind link and abutment truss pin connection rehabilitation
- 2020–2025: suspender rope replacement—planned

Figure 9.20 View from below deck traveler.

Figure 9.21 Temporary class A containment structure used during repainting.

The bridge deck–replacement project of 1987/1988 was performed during the time of increased awareness with respect to the remaining service life of the main cables. The original 7½ in reinforced concrete deck was replaced with a 3 in deep flush-filled steel grid. Precast panels were connected to the existing I beam stringers using all steel haunches. Work was completed at night and accommodated patrons with a single lane of alternating traffic during panel replacement, three lanes during peak rush hours, and two lanes during the balance of the day (see Figure 9.17). Preparation work was performed from below the deck during the day. This new deck reduced the dead load while maintaining or improving the live load capacity from AASHTO H-20 to HS20 (American Association of State Highway and Transportation Officials). The original idea was to make dead-load capacity available for a potential second deck above. This concept for traffic capacity enhancement was abandoned following the main cable evaluation; however, the benefits of the dead-load reduction proved very valuable.

The overlay originally chosen for the concrete-filled steel grid was epoxy mastic with fine stone aggregate broadcast into the epoxy as it is being placed on the deck. This system did not perform well on the MHB. Any minor delamination of the overlay would not allow it to dissipate heat in the hot sun. The epoxy would expand, bubble up, and get destroyed by passing truck tires. Epoxy has a thermal expansion coefficient approximately 10 times higher than that of steel or concrete. This creates a very big challenge for any epoxy/steel adhesive interface.

After an extensive series of repairs and alternative thin overlay trials, the authority decided to abandon the thin chemical overlay solution and go forward with a two-course asphalt system. The first course is 1½ in of the waterproof asphalt called Rosphalt (Royston), an additive combined with a dense sand mix, and the second course is 1½ in of a traditional state-standard asphalt top course. The top course is milled off and replaced on a 12- to 15-year cycle. The waterproofing course has been in place for over 25 years. The dead load saved in the new deck design afforded the authority the option to use a heavier overlay (3 in of asphalt) compared with the thin epoxy overlay.

In summary, 80 years of environmental, dead, and live-load stress have reduced the factor of safety on the MHB main cables from approximately 3.75 to nearly 2.80. In the years since the bridge had its cables opened, inspected, oiled, recompacted, and rewrapped, the authority has developed confidence in the future useful life of these critical bridge elements. However, in the mid-1990s when confidence was low, the authority decided to perform a preliminary design for a cable-replacement project. The existing anchorages were evaluated for reuse and alternative locations for cable anchorages were also explored. Since the existing bridge is a two-cable catenary system, both two-cable and four-cable systems were evaluated. Tower saddle placements were also compared for ease of construction, both above and adjacent to the existing saddles. Load-transfer systems and new suspender rope connections completed the review. This exercise was important to the bondholders when the authority needed to issue a new series of bonds for capital improvements and pledge the revenue from the MHB over the 20-year repayment term.

Bridge component service life and schedules for in-depth inspections and routine maintenance will all vary based upon the bridge environment, traffic volumes, speeds, and vehicle weights. The author hopes that in sharing the history of the MHB service record, other bridge operators may be better prepared to face their challenges in tending to these internationally recognized feats of engineering, also known as world-class suspension bridges.

REFERENCES

1. Federal Highway Administration. National Bridge Inspection Standards, 23 CFR Part 650, Federal Highway Administration, U.S. Department of Transportation, Federal Register, Vol. 69, No. 239, 2004.
2. Guidelines for Inspection and Strength Evaluation of Suspension Bridge Parallel Wire Cables, Report 534, National Cooperative Highway Research Program, Washington, DC, 2004.
3. Mahmoud, K. BTC Method for Evaluation of Remaining Strength and Service Life of Bridge Cables, New York State Department of Transport Report C-07-11, New York State Department of Transport, New York State Bridge Authority, in cooperation with U.S. Department of Transportation's Federal Highway Administration, September 2011.
4. Alampalli, S., and Moreau, W.J. (Eds.). *Inspection, Evaluation and Maintenance of Suspension Bridges*, CRC Press, Boca Raton, FL, 2015.

Chapter 10

Shantou Bay Suspension Bridge

Gongyi Xu

CONTENTS

10.1 BACKGROUND

Guangdong was the pilot province in China for economic reform in the 1980s, and its economic development is among the most advanced. In the early 1990s, Guangdong Province started reform in highway construction and was the first in China to implement the concept of toll roads for finance and management, called as "loans for expressways and tolls to repay the loans." This speeded up the long-planned Guangzhou–Shantou automobile-only highway project and made it a model project for highway construction in China. However, the biggest engineering challenge to this project was to build a major bridge crossing Shantou Bay in order to connect the city of Shantou directly. Therefore, the Shantou Bay Bridge became the critical point that needed to be started first to ensure the success and schedule of the entire project.

Shantou Bay is located to the south of Shantou, facing the East China Sea, and bordering the Taiwan Strait and the western Pacific International Golden Waterway. Shantou Port has been an important trading port on the southeast coast of China since the time of the Ming Dynasty; it played a tremendous role in Shantou's development and prosperity. However, Shantou Bay cut off the north–south highways along the coast of Guangdong and made the travelers rely on ferries to cross the bay. Transportation was very inconvenient without a bridge.

For better implementation of this project, Guangdong Provincial Department of Transportation withdrew the direct management right from the local government, who already performed a study and design for many years on this bay crossing. To find the most economical design that also uses advanced technologies, the department made it a design–build project and invited public bidding across China. After 1 year of tendering and bidding process, Major Bridge Reconnaissance & Design Institute Co., Ltd. (BRDI; the designer) and Major Bridge Engineering Group Co. (MBEC; the contractor) won this bidding for the lowest engineering cost with a suspension bridge design.

10.2 GENERAL DESIGN

10.2.1 Bridge alignment

Before the bidding, China Railway Major Bridge Reconnaissance & Design Institute Co., Ltd. (BRDI), the designer and researcher, made a detailed survey in an area of more than 10 km around Shantou and along the gulf coast. BRDI chose four possible bridge-alignment options for further study in order to determine the most economical and rational alignment for meeting the functional requirements. All the alternative bridge alignments were in the port area, and the navigational clearance governed the span length and height of the main bridge. However, for the same navigation height, the length of the bridge would depend on the water's width and terrains on both sides.

The selected bridge site was located at the open coastal area to the east of Shantou near the mouth of Shantou Bay where the channel is straight. In the middle of the water, there is an island named Mayu, which has relatively high terrains on both sides with good geological conditions. Utilizing the island can reduce the project scale to the minimum size. Meanwhile, it is also the most rational alignment for linking the Shenzhen–Shantou expressway into the national coastal road network.

The selected alignment resulted in no building demolition needs at both ends of the bridge and large spaces for interchanges to connect with the urban roads. This was convenient for not only the passing traffic channels but also meeting the double traffic function of urban transportation and highway traffic during the operation.

10.2.2 Hydrology and navigation channel

Shantou Port is a natural tidal inlet bay and the bridge site is influenced by reciprocating tides. At the bridge site and downstream the long, narrow water channel, the flood strength is 10 m/s and the ebb flow velocity is about 2.0 m/s. The average measured maximum difference between high tides is 2.35 m/s. The average rising tide time is about 7 h and the average falling tide time is about 5.5 h. At the flat shoal water levels, the water in the south of Mayu Island serves large vessels, having an approximate width of 550 m and a maximum depth of 25 m, and the water north of the island is for small fishing vessels, having an approximate width of 350 m and a maximum depth of 8 m. The annual average high tide level is 1.32 m according to the historical hydrological records, which was used as the bridge's navigable water level in design. Because the south channel is restricted by the islands in the sea and the giant reef under water, the waters around the bridge site provide a constrained one-way controlled channel with piloting mainly relying on the two lights located at Shantou bank. The main channel at the bridge site is straight; ship navigations are in good order and concentrated in a deep-water area about 200 m wide.

Considering the above factors, the main channel was designed for 50,000-ton vessel navigation, with a minimum span of the main bridge of not less than 400 m and a vertical clearance height of 46 m.

10.2.3 Geotechnical investigation

The geological exploration report showed that the bridge location is covered with granite, with the bedrock gradually rising from the center of main navigation channel of the seabed to both shores to the south of Mayu Island. Most of the shore rocks are submerged during high tides but appear at low tides. The water depth is about 20 m around Mayu Island north of the main bridge, with a silt covering layer of about 6 to 8 m thick, followed by weathered granite.

To the south of the bridge, the water depth is about 15 m, with a silt-covering layer as well, followed underneath by cohesive clay with a maximum depth of 27 m. The next layer is granite with no obvious weathering evidence but bigger surface undulation. The ultimate compressive strength of the granite bedrock is up to 100 MPa.

In the south landing area of the bridge, the ground is almost covered with exposed granite. It extends to the marine area with reducing thickness from south to north through subnavigation channel and rises to the north landing ground with a maximum depth of 60 m. The seabed is covered with deltaic sedimentary silt, silt sand, and fine sand.

10.2.4 Meteorology

The geographical position of bridge is close to the Tropic of Cancer, where the climate is mild with frequent rainfalls. Weather records showed that in July the highest average temperature is +31.6°C and the lowest average temperature is +25.5°C; in January the highest average temperature is +17.3°C and the lowest average temperature is +10.1°C. In the 35-year period from 1951 to 1985, the highest temperature recorded was +38.6°C in July and the lowest was +0.4°C in January; and the monthly mean relative humidity varied from 82% to 88%.

Near the bridge site, the northeast wind remains dominant nearly all-year round, while southerly winds reign in the summer. Due to the shielding of the Dazhou Island in the east and Guangao Mountain in the south, the wind coming from southeast is usually weak. Typhoon seasons are between May and November every year (mainly concentrated in July–September). When a typhoon lands in the coastal area of Pearl River estuary, it will bring varying impacts to the bridge site. A powerful typhoon landing near the coast of Shantou can bring sustained winds near or exceeding grade 12 with a maximum instantaneous wind speed reaching 55 m/s, which always cause serious damages. According to wind data, the maximum wind speed in the Shantou region is 40 m/s. Wind pressure around the bridge site was determined to be 1200 Pa, corresponding to a basic wind speed of 44 m/s, a 20 m height above ground, a frequency of 1/100, and 10 min for averaging. Considering the influences of geography and terrain conditions, structural characteristics of the bridge, and the importance of wind resistance safety, BRDI determined the basic design wind speed (V_{10}) to be 47 m/s for this bridge, also in reference to international wind design data.

10.2.5 Seismicity

Chaoshan Basin, where the bridge is located, is an earthquake prone region with active seismicity. There have occurred several destructive earthquakes in the past. The earthquake intensity zoning map of China issued by the China Seismic Bureau classifies Shantou as 8 degrees for basic earthquake intensity. According to monographic study data made by the Guangdong Seismological Bureau, the bridge site's probability seismic intensity was determined to be 8 degrees, with $P = 0.1$, $T = 100$ years, and $k = 0.2229$ as the maximum lateral seismic factor of the bedrock ground in design.

10.3 SUSPENSION BRIDGE DESIGN

Based on the site conditions described above, the bridge layout was established as shown in Figure 10.1. The main channel crossing is a three-span, two-hinge suspension bridge of span lengths 154 m + 452 m + 154 m.

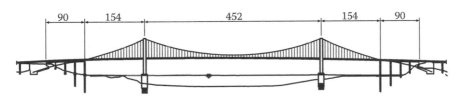

Figure 10.1 Main bridge layout elevation.

10.3.1 Support system

The stiffening girder of the bridge was made of a prestressed concrete box section with three independent spans of 150, 444, and 150 m. Based on the mechanical characteristics of the structural system, pulling and pushing bearings were arranged at both two ends of each span. There is a transition girder at the pylon location between the side span and the main span, and transverse wind-resistance bearings were used at both ends of the middle cross frame of the transition girder, as shown in Figure 10.2.

The transition girder adopted a composite structure design that consisted of a steel truss and a concrete slab, as shown in Figure 10.3.

The design intent was to produce a special structure that can fully transfer the longitudinal axial forces and displacements, with certain stiffness in the transverse direction, and simultaneously with little bending stiffness in the vertical direction. A three-span, two-hinge stiffening girder support system was determined to be the optimal solution.

10.3.1.1 Vertical supports

Different from the pendulum structure adopted in other suspension bridges, vertical pulling and pushing bearings were utilized as vertical supports. This choice was based on the considerations for releasing the horizontal end displacements of the girder due to factors such as concrete shrinkage and creep under prestress. Two brackets were respectively located at both

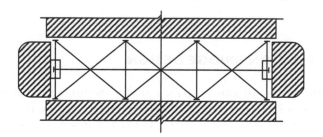

Figure 10.2 Transition steel truss plan view.

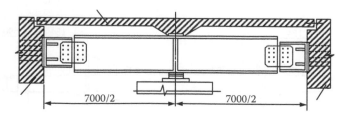

Figure 10.3 Transition girder elevation view.

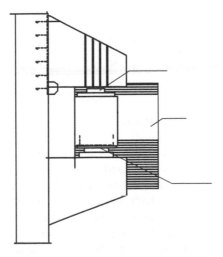

Figure 10.4 Vertical support.

ends of the side piers and on both sides of the pylons, and spherical rubber supports were set on the contrapositive brackets at the girder ends for clamping purpose to limit the vertical displacements. Figure 10.4 demonstrates the working principle of the bearings. Practice has proved that this structure can properly accommodate different end deformations caused by uncertain influences, and it can provide vertical restraint without limiting the rotational and longitudinal displacements.

10.3.1.2 Lateral supports

Lateral flexural rigidity can be provided by a wind-resistant truss on the same plane as the transition girder. This system forms a lateral continuous structure as shown previously to reduce lateral deflections of the girder under wind actions. To achieve the goals of relatively independent force condition for the three spans of concrete stiffening girder and making posttensioning operation convenient, the lateral bearings were respectively arranged on the inner side of the pylons and the top of the side-span piers.

10.3.2 Stiffening girder

The prestressed concrete stiffening girder has a single-box, triple-chamber cross section, as shown in Figure 10.5. The central height of the girder is 2.20 m, and the side webs' height is 0.96 m; the top slab of the girder has a width of 24.20 m and crown-shaped 2% cross slops. The bottom slab of the girder is of a circular arc with a 76.88 m radius and a 24.72 m horizontal width. The side surfaces are inclined to reduce wind drags. The sling points on the two sides of the girder are connected through a solid diaphragm

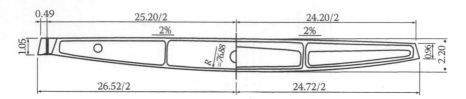

Figure 10.5 Stiffening girder cross section.

across the box section, which forms a lateral stress-passing system. The middle chamber of the stiffening girder has a width of 8.82 m, and the width of the side chambers is 7.25 m each. The thickness of the interior webs is 0.18 m; the side webs have a thickness varying from 0.26 m at the top to 0.52 m at the bottom, which provides a space for pre buried adapting pieces during construction and strengthen the corners.

The top slab of the stiffening girder has a thickness of 0.18 m for wheel loads, and the thickness of the bottom slab is 0.14 m. The thickness of the main diaphragm is 0.18 m across the middle chamber and increases to 0.6 m across the side chambers and the sling points with fluent transitions. The diaphragms have a longitudinal spacing of 6 m at the sling points; and there is a subsidiary ribbed diaphragm at each wet construction joint (closure pour) that has a thickness of 0.3 m to reduce the influence of manufacturing errors and reduce the longitudinal span lengths of the slabs for local stability concerns.

Layout of prefabricated segments of the stiffening girder requires an overall consideration of factors such as the mechanical characteristics and construction conditions. All the 5.7 m long girder segments, each centered at the sling points, were prefabricated near the bridge site. There is a construction joint of 0.3 m width between any two adjacent segments, and these joints are casted with a hollow subsidiary cross beam after the segments are in place. The main span of the stiffening girder has 73 segments and each of the two side spans has 24 segments, for a total of 121 segments across the entire bridge. Four prefabrication stations were set up near the bridge site, where all the segments were casted and partially posttensioned with the transverse tendons. The segments were then transported to a storage area through temporary supports and stored for 6 months per design requirements for eliminating the impact of shrinkage and creep. Finally, the segments were transported by barges directly under the bridge for vertical lifting and transferring to the suspenders after the erection sequence. A maximum of four segments can be installed in 1 day, with each weighing 170 tons.

10.3.3 Posttensioning of stiffening girder

The longitudinal posttensioning system was designed to supply the full prestressing needs to the stiffening girder for its vertical bending moment envelope caused by dead and live loads, as illustrated in Figure 10.6.

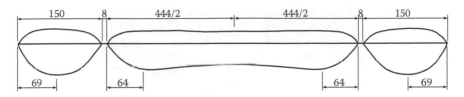

| 150 | 8 | 444/2 | 444/2 | 8 | 150 |

| 69 | 64 | | 64 | 69 |

Figure 10.6 Vertical bending moment envelope of stiffening girder.

The longitudinal tendons were arranged based on the central alignment principle, which means that they bring only axial force to the girder section. The high-strength, low-relaxation 7–15.2 mm steel strand was adopted for the longitudinal tendon, which is composed of seven 15.00 mm diameter wires with a specified tensile strength of 1670 MPa per strand. There are 127 strands in the girder cross section in the side span where the bending moment is the highest, of which 64 strands are placed in the top slab and 63 strands in the bottom slab. The maximum moment cross section in the main span has 109 strands: 54 in the top slab and 55 in the bottom slab. As shown in Figure 10.7, the strands were arranged based on the bending moment demand, with a minimum lateral spacing of 0.3 m, and gradually anchored at the solid diaphragms. Due to the different thicknesses of the top and bottom slabs, the top tendons were internal and bonded, while the bottom tendons were external and unbonded with high-density polyethylene sheathings to reduce prestress loss caused by concrete shrinkage and creep and provide measures for future prestress replenishing if needed. The bottom prestress strands are shown in Figure 10.8.

In the transverse direction, the stiffening girder was partially prestressed considering the participation of reinforcing steel bars. There were two types of lateral posttensioning tendons. One was bonded for heavy loading with each strand consisting of nine 15 mm diameter wires and four strands in each standard precast segment distributed in the main diaphragm chamfer; and they were prestressed at both ends through the sling points. The other type of tendon was unbonded 15 mm diameter single wires; and there were 24 strands in each standard precast segment distributed in the chamfer on both sides of the main diaphragm: 6 near each bottom corner with a 0.08 m spacing, 5 on each side away from the main diaphragm spaced at 0.04 m, and the remaining 2 in the two cast-in-place subsidiary diaphragms tensioned symmetrically.

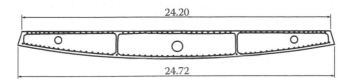

24.20

24.72

Figure 10.7 Arrangement of posttensioning strands.

Figure 10.8 The bottom prestress strands in a stiffening girder.

10.3.4 Seismic design of stiffening girder

Because of the critical wind condition at the site, the bridge adopted a con-crete stiffening girder that has higher wind resistance. However, the increase in self-weight brought some adverse effects. Some special design ideas aim-ing at this contradictory provided three measures for seismic safety. The first was elastic constraints from special steel strands connected to the pylons. When the girder vibrates in an earthquake, the strands will be stretched, possibly to fracture; thus, part of the seismic energy can be released and excessive girder displacements can be limited. Secondly, tailor-made rubber cushions were installed under the expansion joints at both ends of the three-span girder system to absorb the displacements and the inertia forces caused by seismic events. The final line of defense is for a rare earthquake: a shear wall was designed based on the girder displacement in a condition after both the overloaded transition girder and the expansion joints have been destroyed. This multilevel defense system can enhance the seismic safety of the main structure, including the pylons and the stiffening girder.

10.3.5 Pylons

The pylons are three-strut portal frames made of C50 (28d yield stress: 50 MPa) reinforced concrete. The height of the pylon above the pile cap is 95.10 m, and each column is hollow D shaped with external dimensions of 6.0×3.5 m². For wind resistance, three rectangular prestressed concrete beams connect between the two columns, as shown in Figure 10.9.

10.3.6 Pylon foundations

The pylon foundations were made of detached caisson-enclosed grouped piles at the upstream and downstream sides. This design solved the problems

Figure 10.9 Pylons of Shantou Bay Bridge.

caused by complex geotechnical conditions, such as elevation differentials and various weathering conditions of subgrade strata, thus ensuring the convenience and safety of piling construction. Each pile group consists of six piles enclosed by a concrete-filled steel caisson to form a composite system. The grouped piles connect with the pile cap beam to form a portal frame structure that can reduce the free length of the piles, as shown in Figure 10.10.

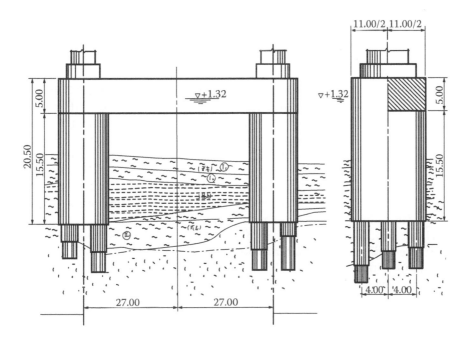

Figure 10.10 Foundation of main bridge pylons.

Figure 10.11 The south anchorage.

10.3.7 Anchorages

Gravity-type rock-socked anchorages with embedded steel structures were used to anchor the main cables. Mountain rocks were utilized to compose a rock-embedded structure and extend the main cable for 95 m to the anchorages. This system saved the excavation volume and the quantity of concrete for constructing the anchorages. The south side anchorage is shown in Figure 10.11.

10.3.8 Main cables

The spacing between the centerlines of the two main cables is 25.2 m, and the PPWS method was employed to erect the cables. There are 110 strands in each cable, with each strand consisting of 91 of 5.1 mm diameter wires. Both ends of each strand group were installed with a hot-cast saddle that was connected to the embedded anchor rod in the anchorage. The sag-to-span ratio in the main span was 1:10, which resulted in a 1:29.6 ratio in the side spans. A swing column was employed at the top of each end pier to support the cable vertically to improve the loading condition of the stiffening girder in the side spans, as shown in Figure 10.12.

The suspenders were comprised of four wire ropes, which were connected up to the main cable through a grooved cable band and down to the stiffening girder through a special connector, as shown in Figure 10.13. The connector served two purposes: (1) connect between the four wire ropes of the suspender and the two rods from the stiffener girder and (2) provide length adjustments to the suspender. The suspenders were made of galvanized steel wire ropes with a 45 mm outer diameter and a 1700 MPa tensile strength, with a design safety factor of 3.5.

Figure 10.12 Side supports for the main cable.

Figure 10.13 Main cable and the hanger ropes.

Figure 10.14 Splay saddle at anchorage.

10.3.9 Cable saddles

The cable saddle on top of the pylon was made of cast steel and composed of the upper saddle and the nether slab. For the convenience of transport and lifting, the upper saddle was divided into two parts in the longitudinal direction with each weighing 20 tons. The upper saddle was connected to the nether slab through high-strength bolts after being set in the right position.

The contact surfaces between the upper saddle and the nether slab were specially treated for rust protection and friction reduction before leaving the factory to ensure smooth movements during the positioning process in erection.

The splay saddle at the anchorage was also made of cast steel with each weighing 19 tons. As shown in Figure 10.14, there is a special sliding pot bearing under the saddle to help adjust the deformations of the main cable and the anchor rods.

10.3.10 Quantities of suspension spans

- Steel weight of cable wires: 3180 tons
- Concrete volume of anchorages: 29,800 m^3
- Concrete volume of pylon foundations: 9600 m^3
- Concrete volume of pylons: 3500 m^3
- Concrete volume of stiffening girder: 9345 m^3

10.4 CONSTRUCTION OF SUSPENSION BRIDGE

10.4.1 Stiffening girder prefabrication

Segments of the stiffening girder were prefabricated in a yard located on the south shore of Shantou Bay. Four casting stations were used to fabricate all the girder segments. The bottom dimensions of all the casting molds remain unchanged through the fabrication process, while the end and internal dimensions were adjusted for different segments per design drawings. The near-site prefabrication was implemented following standard procedures to assure quality. The four casting stations generally produced four to five girder segments per month.

Special considerations were given to the stresses the girder segments may experience while in storage and during lifting. Special lifting equipment and a supporting system were designed and used to ensure the same stress state the segments will experience in their permanent position. This not only avoided additional strengthening or reinforcing needs for the prefabricated segments but also made full tensioning of the transverse tendons possible at the casting station, thus eliminating unnecessary aboveground operations.

Considering the long-term storage and the large area occupied in the fabrication yard, the prefabricated segments were stored in double-layer stacks, as shown in Figure 10.15.

10.4.2 Stiffening girder erection

Erection of the stiffening girder strictly followed the design requirements. The operation proceeded simultaneously from both the north and south ends of the bridge. As the technology used for friction reduction on the

Figure 10.15 Prefabricated girder segments in storage yard.

main saddle was new, no previous data were available. In order to prevent problems caused by shifting at the main saddle during the girder-lifting process, an asymmetric lifting scheme was adopted based on design calculations. The process began with erecting the first 12 segments on the center-span side of each pylon, as shown in Figure 10.16. Then the increased

Figure 10.16 Erection of stiffening girder—at beginning.

Figure 10.17 Floating and hoisting the segment.

Figure 10.18 Erection of stiffener girder—near completion.

horizontal tension in the main cable on the center-span side was gradually balanced by adjusting the side-span cable sag. When the offset at the main tower saddle was completely pushed back to the permanent position, based on the balance requirement of the main cable horizontal tension, subsequent stages of girder segment–erection sequence were determined, and then began the synchronous, symmetrical girder erection from both sides of the pylon. Figures 10.17 and 10.18 depict two different stages in the stiffening girder–erection process.

10.4.3 Temporary connections and cast-in-place joints

Temporary connections between precast deck segments were made by extruding steel sections embedded in the side webs at both ends of the girder segment to ensure that both the alignment and the gap between abutting segments meet the design requirements, as shown in Figure 10.19. After all the girder segments were hoisted in place, sandbags were placed on the girder to simulate part of the second-stage dead load, and girder alignment was adjusted along the entire bridge according to design requirements. At this time, bolts connecting the steel sections were gradually fastened along the bridge to form the final locked position. Cast-in-place joints between segments were then poured with concrete from the span center toward both sides symmetrically, making the stiffening girder continuous along the bridge. Figure 10.20 shows a cast-in-place joint just before pouring of concrete.

10.4.4 Longitudinal posttensioning

After the cast-in-place joints had been cured, an independent continuous concrete beam formed in each of the side spans and in the main span.

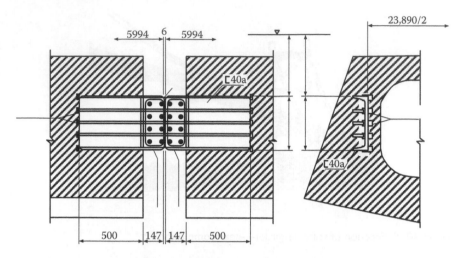

Figure 10.19 A temporary connection between precast girder segments.

Figure 10.20 A cast-in-place joint before pouring of concrete.

Longitudinal prestressing tendons were then installed and tensioned in each of the continuous units. The design specified synchronous and symmetric tensioning between the top and bottom slabs and between the left and right webs. The design also required long tendons to be tensioned first and short tendons afterward within each span.

10.4.5 Installation of transition beams

Transition beams should be installed after the posttensioning of the stiffening girder in all spans and after the application of deck dead loads. This can avoid secondary stresses due to posttension-induced elastic compression and the effects of second-stage dead loads. The transition beams were installed in two steps: first, completing the adjustments and connections of steel trusses and, second, pouring concrete to form a structural hinge in the continuous deck.

10.4.6 Construction of pylon foundations

Pylon foundations were constructed in the following sequence: hoisting the steel caisson in place, as shown in Figure 10.21; sinking the caisson into required depth by using weights made of water tanks; removing silt from inside the caisson by using pumps, placing tremie concrete to seal the bottom; drying the caisson by pumping water out, drilling and casting rock-socketed piles; and finally, filling the caisson with concrete to make the pile group an integrated system. This type of foundation was found to be economical and fast to construct, creating a successful example for bridge foundations in the sea.

Figure 10.21 Hoisting the steel caisson of pylon foundation.

Figure 10.22 Suspension system erection.

10.4.7 Erection of suspension system

The suspension system included the main cables, saddles, cable bands, clamps, suspenders, and so on. The conventional construction methods for modern suspension bridges were employed for this project, as illustrated in Figure 10.22.

10.4.8 Load testing

After completion of construction and before opening to traffic, the bridge was tested under dynamic and static loads and checked against acceptance requirements. Those tests verified that the engineering quality met the design requirements and also established original records for bridge maintenance in the future. The test results agreed with theoretical calculations and met the acceptance requirements. Figure 10.23 depicts load testing of the Shantou Bay Bridge.

Figure 10.23 Bridge load test before opening.

10.5 SUSPENSION BRIDGE OPERATION MANAGEMENT

Shantou Bay Bridge opened to traffic on December 28, 1995. Figure 10.24 is an overview of the main bridge.

A company was established by shareholders who participated in the bridge investment, responsible for operational management and bridge-maintenance work.

Since Shantou Bay Bridge was the first modern suspension bridge in China, no domestic maintenance management experience was available for reference. Thus, a technical advisory committee was formed (Figure 10.25); this consisted of five experts (Figure 10.26): Bingsen Gong, the chief

Figure 10.24 Shantou Bay Bridge opened in 1995.

Figure 10.25 Bridge inspection by advisory committee.

Figure 10.26 Advisory committee meeting. Left to right: Bingsen Gong, Jin Yang, Xiangang Hu, Guobin Shi, and Gongyi Xu.

construction engineer; Professor Jin Yang, the chief design engineer; Guobin Shi, the chief supervision engineer; Xiangang Hu, the owner's project manager; and the author, Dr. Gongyi Xu, the chief designer of the suspension bridge. These five committee members met once or twice a year, providing consultations on various problems related to bridge operation and maintenance. These make the maintenance works have a very high efficiency.

Based on a recommendation by the advisory committee, the "Shantou Bay Bridge Maintenance and Repair Manual" was developed. The manual describes the main principles and methods for long-term bridge monitoring and regular inspections, as well as the requirements for selecting professional companies through competitive bidding for bridge inspections and maintenance. The advisory committee reviews the long-term monitoring and annual bridge inspection reports, reviews and advises on bridge-maintenance projects and repair plans for the next year, and recommends actions to be implemented by the bridge-management company. This model has proved to be simple, efficient, focused, and successful from its 20 years of implementation history.

10.5.1 Major maintenance projects

10.5.1.1 Replacement of pavement in suspension spans

The original pavement design for the stiffening girder of the suspension bridge was 6 cm of waterproof steel fiber–reinforced concrete. After less than 2 years of operation, the pavement experienced map cracking, as shown in Figure 10.27. In order to prevent concrete spalling for traffic safety, the original pavement was replaced with asphalt concrete, as shown in Figure 10.28. The replacement asphalt pavement has

Figure 10.27 Cracking on concrete pavement.

Figure 10.28 Replacement of pavement with asphalt.

performed well for 15 years, which has already exceeded its expected service life.

10.5.1.2 Protection of main cables

The original protection of the main cables included a corrosion-inhibiting paste on the steel wire surface, wrapping with galvanized steel wire, a two-coat paint, and an antiskid coating on the top surface. After 5 years, the coating surface experienced powdering and cracking, and repairs were made, as shown in Figures 10.29 and 10.30.

Repair of the cable-protection system included removing the old multi-layer paint outside the wrapped wire and painting with a new polysulfide compound. This compound became a 3 mm thick rubber sleeve tightly

Figure 10.29 Applying corrosion-inhibiting paste on the main cable.

Figure 10.30 Painting on the main cable.

Figure 10.31 Renewed main cable protection.

wrapping around the main cable after vulcanization, providing waterproofing and corrosion protections. Another antiskid coating was then applied to the top surface of the main cable. The service life of the new coating was estimated to be about 30 years or more. Figure 10.31 shows the renewed main cable protection.

10.5.1.3 Suspenders

After 18 years of operation, the owner considered to replace all the suspenders due to concerns that ocean climate–induced corrosion may reduce their load-carrying capacity, thus not meeting the safety requirements. Figure 10.32 shows some signs of corrosion experienced by a suspender.

However, some experts did not agree to replace all the suspenders on the bridge. Ultimately, eight relatively seriously corroded suspenders were replaced as a compromise. Breaking load tests were performed on the removed suspenders to examine their actual capacities, and test results are shown in Table 10.1.

It can be seen from Table 10.1 that after 18 years of outdoor use, the most seriously corroded suspender had a breaking capacity 11.7% lower than the design value. These indicated that the suspenders can meet the safety operation requirements and do not need to be replaced. Figure 10.33 shows painting operation on a suspender.

Figure 10.32 Inspection of hanger ropes.

Table 10.1 Breaking test results of eight suspenders removed from bridge

Hanger sample	Actual breaking force (kN)	Breaking location	Loss (%)
MS26-S	1259	–	7.2
MS26-N	1198	Near socket	11.7
MN32-S	1429	Near socket	–
MN32-N	1304	Near socket	3.8
MN11-N	1388	Far socket	–
MN11-S	1282	Near socket	5.5
S18-N	1419	Near socket	–
S18-S	1270	Near socket	6.3

Note: The design breaking force for all hanger samples was 1356 kN.

Figure 10.33 Repainting of suspender hanger ropes.

10.5.1.4 Expansion joints

The large-movement, rail-type expansion joints at both ends of the suspension bridge were developed and manufactured in China. They were designed for rare earthquakes, with a maximum movement range of 0.80 m. They have performed satisfactorily for 20 years, with the rubber sealing strips replaced only once.

10.5.1.5 Water seepage control in anchorage rooms

Different levels of water seepage occurred in all four anchorage rooms after the first rain reason after the bridge opening. Antiseepage measures were taken and dehumidifiers were installed in the anchorage rooms in the subsequent year. The humidity level in the anchorage rooms has been controlled to be 40% below the design requirements.

10.5.2 Other issues

Overweight vehicles are a common problem in China, despite the general recognition that they can cause fatal damages to bridge structures. Statistics show that more than 80% of truck or axle loads of freight vehicles exceed the legal levels allowed by transportation standards. Weigh stations were recently constructed at the bridge and are expected to come into effect in 2015.

Due to age and health concerns, some members of the advisory committee will be unable to provide continued support to Shantou Bay Bridge. The owner will need to revise the original model for expert consultation.

10.5.3 Maintenance plans

10.5.3.1 Overlay renewal

The existing roadway wearing surface has exceeded its expected service life, with its thickness badly worn, waterproof layer damaged, and surface condition not meeting the requirements. It will need to be replaced completely in the near future.

10.5.3.2 Main cable protection

In recent years, wire corrosion has been observed inside the main cables. In order to prevent continued deterioration of the cables, dehumidification systems will be installed to blow dry air inside the main cable, as shown in Figures 10.34 and 10.35.

10.6 BRIDGE TRAFFIC

Vehicle traffic volume on the Shantou Bay Bridge (Figures 10.36 and 10.37) increases year by year. In 2014, vehicle volume was the highest since the bridge opened in 1995, with a yearly vehicle number of 7,944,516, including

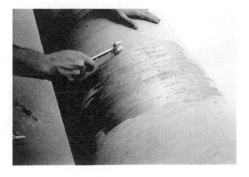

Figure 10.34 Improving the life of the cable.

Figure 10.35 Main cable protection with dehumidification.

Figure 10.36 Shantou Bay Bridge.

Figure 10.37 Another view of Shantou Bay Bridge.

5,807,417 cars and 2,137,099 trucks. The maximum daily vehicle volume occurred on February 22, 2015, for 55,412 in total, including 40,506 cars and 14,906 trucks. Statistical data from January to December 2014 indicated that the trucks accounted for about 26.9% of total traffic, and 10.7% are overloaded trucks weighing more than 55 tons.

10.7 SUMMARY

Construction of the Shantou Bay Bridge began in March 1992, and the bridge opened to traffic in December 1995. Prestressed concrete boxes were selected for this suspension bridge to resist strong, hard wind conditions. With a prestressed concrete stiffening girder as its main feature, the design and construction of this suspension bridge provided valuable experience for long-span structures of this type. The Shantou Bay Bridge set new world records for the thinnest wall thickness, longest prestressing tendons, and the longest span length for prestressed concrete box girders. Taking advantage of the low costs in labor and concrete material in China, the innovative design achieved the best wind stability for its long spans and demonstrated the technical wisdom a bridge engineer should possess.

As the first suspension bridge in China, the success of the Shantou Bay Bridge set a new milestone in the modern bridge history. Besides its technical values in design, fabrication, construction, management, and wind tunnel test, it also promoted the applications of several new industrial products including galvanized high-strength steel wires, suspender ropes, large-scale cast-steel elements, and cable corrosion–protection materials. These industries expanded rapidly in China after this project and have become internationally competitive. Several long-span suspension bridges were constructed after the 760 m long Shantou Bay Bridge, including the 900 m long Xiling Yangtze River Bridge, the 888 m long Humen Bridge, the 1385 m long Jiangyin Yangtze River Bridge, and the 648 m long Xiamen Haicang Bridge. Today, with the recent start of construction of the 1700 m long double-deck suspension Yangsigang Yangtze River Bridge in Wuhan, China has become the country that has the largest number of suspension bridges in the world.

ACKNOWLEDGMENTS

This chapter would not have been possible without the great help I received from many individuals mentioned here. I was the last contributing author invited for the book, thanks to a recommendation to Dr. Sreenivas Alampalli by Dr. Ed Zhou. As a result, I also had a very short period of time to prepare this chapter. Dr. Ed Zhou, the national practice leader of Bridge Instrumentation and Evaluation of URS, also made great efforts in editing the chapter. Dr. Sreenivas Alampalli offered great encouragement and a grace period for the submission deadline. And then thanks to my wife, Liang Zhihong, for always serving me.

I wish to acknowledge many of my colleagues at BRDI of China Railway: Liao Mujie, Dr. Xu Rundong, Liu Xiaolin, Sheng Changfang, Zhou Jie, Wei Yushan, and Du Fang. I also owe a debt of gratitude to the bridge owner, Shantou Bay Bridge Co., Ltd., for supplying the information on bridge management and maintenance. Last but not the least, I wish to express my appreciations to my company, BRDI; the Bridge Science Research Institute of MBEC; the Railway Engineering Research Institute of Chinese Academy of Railway Sciences; and the Center for Bridge Inspection and Maintenance Technologies of BRDI.

Chapter 11

Kingston–Port Ewen Bridge

Sreenivas Alampalli

CONTENTS

11.1 INTRODUCTION

The Kingston–Port Ewen Bridge (Figures 11.1 through 11.3) over Rondout Creek, as well as the New York State Barge Canal and a city street, is the only suspension bridge owned and maintained by the New York State Department of Transportation (NYSDOT) and was opened in 1921. The bridge is currently known as Wurts Street Bridge as it carries Wurts Street, but it is also sometimes referred to as Rondout Creek Bridge or Old Bridge. It is located in the city of Kingston and crosses over an island in the Rondout Creek known as Island Dock (actually an artificial island created in the days of the Delaware and Hudson Canal and once home to an active shipbuilding facility), the New York State Barge Canal, and a city street known as Dock Street. The City of Kingston is located about 100 mi north of the New York City, which is home to well-known suspension bridges, such as the Brooklyn Bridge, the Manhattan Bridge, and the Williamsburg Bridge. Although NYSDOT conducts the biennial inspection of the Manhattan, Brooklyn, and Williamsburg suspension bridges, the

275

Figure 11.1 Kingston–Port Ewen Bridge.

Figure 11.2 Another view of Kingston–Port Ewen Bridge.

New York City Department of Transportation is the primary owner and is responsible for upkeep of these bridges.

As noted above, the bridge carries Wurts Street, a city street, with an average daily traffic of approximately 5000 as per 2013 estimates. The bridge is currently weight restricted to vehicles weighing less than 3 U.S. tons; thus, trucks are not allowed on the bridge (see Figure 11.4). The bridge deck

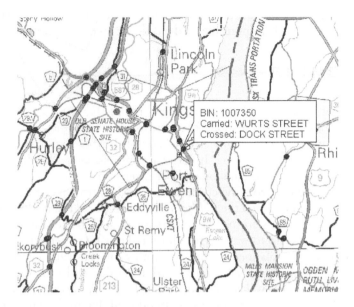

BIN: 1007350
Carried: WURTS STREET
Crossed: DOCK STREET

Figure 11.3 Kingston–Port Ewen Bridge location.

Figure 11.4   Posting sign at the bridge.

supports two lanes of traffic, one lane in each direction. NYSDOT is the primary owner and is responsible for bridge maintenance and operations.

11.2 HISTORY

Wurts Street Bridge was built in 1921 to complete New York's first north–south automobile highway along the Hudson River's west shore and was considered an important engineering accomplishment associated with the development of early motoring. It was a prominent visual landmark, connecting Kingston and its neighbors of Port Ewen across the creek. Previously those who wished to cross the creek at the south entrance to Kingston had to use a chain ferry named the *Skillypot*. It was reported that *Skillypot* had been able to carry only from 8 to 10 cars a trip and in the summer season it was often necessary to wait up to 3 h on either side to get across [1–10].

Although it was planned in the beginning of the 19th century, construction on the bridge was put off until 1916 due to local political and financial matters, as well as design issues. In 1919, several designs were considered (see Figure 11.5) and then abandoned due to design issues during substructure construction. Construction began again in 1920. The cornerstone of the bridge was officially set by Governor Alfred Smith in 1920 (see Figure 11.6). The construction of the bridge took approximately 1 year, during which the contractors employed a woman as a welder—commonplace during World War II but unheard of in 1920 [1–10].

Ten thousand people attended the bridge's dedication on November 2, 1921, by Governor Miller. The bridge carried State Route 9W until 1977 when a new bridge was built to the east and Route 9W was relocated out of the center of Kingston (see Figure 11.7). The original bridge still stands to reflect the state-of-the-art engineering skills of the early 1900s. Construction of the Rondout Creek Bridge was considered very important, even though under current standards it may be a relatively small suspension span, since it demonstrated the adaptability and economic viability of suspension bridge design and construction principles to relatively short-span highway bridges. The reported cost of the bridge immediately after the construction was $700,000 [1–10].

In 1980, the bridge was listed on the National Register of Historic Places (90NR01106) and represents a key development of early motoring. The riveted-steel cable suspension bridge exhibits the foremost technology and design features of its time and the structure has achieved significance as a prominent visual landmark in its Kingston–Roundout Creek setting. Well-known bridge designer D. B. Steinman stated that "the bridge was the second bridge in the world and the first in the United States to have a hingless, continuous stiffening truss and was the first suspension bridge to have the towers with their main legs battered, providing more clearance for the roadway and increased stability." The bridge is also unique as there are

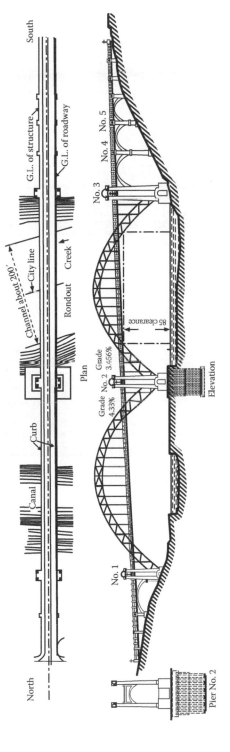

Figure 11.5 Truss design considered before suspension bridge. (From "Rondout Creek Bridge Design Condemned for Excessive Load on Batter Piles," *Engineering News Record*, 88, no. 7, 329–332, August 14, 1919.)

Figure 11.6 Cornerstone of the bridge set in 1920.

Figure 11.7 The new bridge parallel to Wurts Street Bridge.

more strands in the suspension cables on the approach spans than on the main span [11,12].

The north end of the bridge abuts the National Register of Historic Places' West Strand Historic District (90NR01103), a 57-acre area with 258 residential, public, and commercial structures depicting the 19th century. Thus, given the historical significance of the bridge and the surrounding area, any future work on the bridge must be coordinated with the New York State Office of Parks, Recreation, and Historic Preservation [12].

11.3 BRIDGE DESIGN AND CONSTRUCTION

The Wurts Street Bridge is a three-span steel structure with a main span of 705 ft and side spans each of 220 ft. The deck consists of precast concrete planks with an asphalt concrete wearing surface and steel railings. The total bridge width is about 37.3 ft with a curb-to-curb width of 23.2 ft to support the approach width of about 24 ft. The roadway is located 85 ft above the creek. Two 10.5 in main cables carry the two-lane roadway with sidewalks on both sides on a riveted-steel stiffening truss supported by steel towers and piers. Figure 11.8 shows the bridge during the construction [11].

The bridge was built for the New York State Highway Commission. The plans and specifications were prepared by Daniel E. Moran and William H. Yates, consulting engineers from New York City (see Figure 11.9). The bridge was built by the general contractor Terry & Trench, Inc., of New York City. Charles Michaud of Kingston was the subcontractor for the concrete piers and anchorages. W. E. Joyce was the resident engineer for the State Highway Commission and M. Bebarfald was the resident engineer for Terry & Trench. All cable wires used in fabricating the main cables, including wire ropes used for temporary footbridge and suspenders, was

Figure 11.8 Bridge during the construction. (From Joyce, W.E., and M. Bebarfald, "Building the Rondout Creek Highway Suspension Bridge," *Engineering News Record*, 89, no. 11, 424–428, September 14, 1922.)

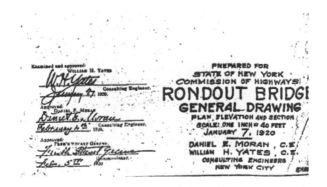

Figure 11.9 Plan sheet showing the names of the designers.

furnished by the John A. Roebling's Sons Company of Trenton, New Jersey. The reported weight of cables and suspenders was 300 tons with structural steel of towers, trusses, and floor system weighing 1430 tons [13].

The bridge was designed for the 20-ton truck and 80 lb/ft^2 lane loading that was standard highway loading at the time of design. The stiffening truss is continuous over the three spans with 17 ft 7.5 in panel length, while the truss depth varied from 15 ft at the tower to 10 ft at the midspan and ends. The trusses and the deck are supported by 1.25 in wire rope suspenders looped over the cables and connected by adjustable U-bolts to the truss verticals [11].

In the main span, each of the two 9.75 in diameter cable contains 1974 galvanized wires of number 6 Roebling gauge. In the side spans, 152 wires were added to each cable to account for the additional stresses from the slope of the short end spans. Hence, cable diameter increased to 10.25 in. Cables were directly anchored into rock anchorages.

Two 152 ft tall steel towers support the structure that rests on concrete piers. For smaller suspension bridges, the towers are normally erected by gin pole or by stationary derrick alongside. For the Wurts Street Bridge, a guyed derrick with 95 ft steel boom was set up on an 80 ft high square

Figure 11.10 Steel towers erected by derricks set on wooden towers. (From Joyce, W.E., and M. Bebarfald, "Building the Rondout Creek Highway Suspension Bridge," *Engineering News Record*, 89, no. 11, 424–428, September 14, 1922.)

timber tower for the erection of each steel tower (see Figure 11.10). The same derricks later erected the adjoining panels of the stiffening truss. The south pier is approximately 18 ft higher than the north pier to accommodate the 5% deck grade from the north anchorage to the center of the bridge and then on 0.2% grade to the south end [11,14].

Anchor pits were 6 ft diameter shafts cut into rock to a depth of about 60 ft below ground surface (see Figure 11.11). They widen at the bottom into a 6×14 ft² chamber of 6 ft height. Each of the two cables has independent anchor pits. The eyebar anchors are embedded in concrete that fills the shafts, but their upper ends are exposed in chambers formed below the

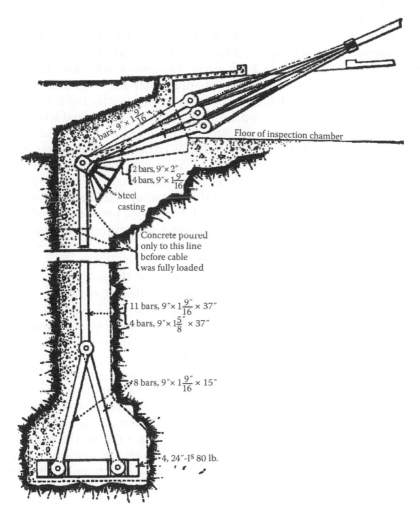

Figure 11.11 Anchorage construction. (From Joyce, W.E., and M. Bebarfald, "Building the Rondout Creek Highway Suspension Bridge," *Engineering News Record*, 89, no. 11, 424–428, September 14, 1922.)

roadway level so that attachment of the cable strands to the anchors is open to inspection [11].

Each anchor was built of four 12 ft long, 24 in deep I beams, connected by a chain of eyebars to the cable shoes. The north tower was founded on two independent 11×16 ft^2 concrete piers, one under each leg. At the south tower an old pier built under a former contract was used with the addition of two 8 ft high pedestals of size 16×16.5 ft^2. The towers had two legs of box section, battered inward from a spacing of 38 ft at base to 27 ft at the top, anchored to the pier by four 2.5 in anchor bolts per leg. In fabrication, legs were built in five sections each. The maximum section weighed 22 tons. Figure 11.12 gives details of plant arrangement for construction of the bridge.

When all the wires were strung and temporary strands were removed, petroleum grease was slushed between the individual wires, and the cable was squeezed by a hand squeezer to the required diameter and again held by wire seizing spaced about 2.5 ft apart. The cable bands were placed and bolted in position. Suspender ropes of appropriate size were socketed and placed in position for stiffening truss attachment. After the full load was on the main cables, they were wrapped with number 9 galvanized wire. Once the construction was completed, all structural steel was painted with a field coat of red lead paint and two finishing coats of battleship gray. Cable spinning operations and more details of construction can be found in the literature described in detail by the engineers in charge of the project [11].

The bridge not only demonstrated the economic viability of suspension bridges almost a century ago for medium-size spans, but it also displayed a number of unique design features that distinguish it from other suspension bridges [15]:

- The stiffening truss is unique in that it was the first American suspension bridge to employ a continuous stiffening truss (it was only the second major bridge in the world to do so, the Cologne Bridge in Germany being the first). This resulted in a more rigid structure for the same weight of material and eliminated the need for expansion joints at the towers (see Figure 11.13).
- The main cable is unique in that it is one of only a few parallel-wire bridges to utilize auxiliary strands in the backstays (as part of the original construction) to compensate for the higher stress resulting from the shortness of the side spans. The auxiliary strand was sized to result in approximately equal stress in the cable on the main-span and side-span sides of the tower, even though the side spans have a much higher angle of inclination (the side spans are only one-fourth the length of the main span, while the ratio is normally closer to 1/2). This auxiliary strand, consisting of 76 wires, was looped around a strand shoe attached to the saddle, thus contributing 152 additional wires to the cross section of each backstay (2126 total wires versus 1974 wires in the main span).

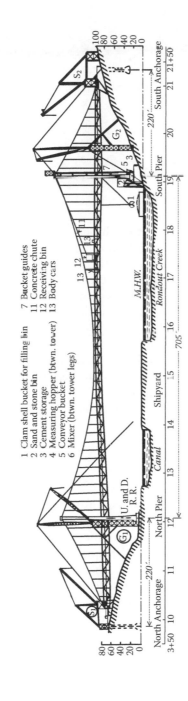

1 Clam shell bucket for filling bin
2 Sand and stone bin
3 Cement storage
4 Measuring hopper (btwn. tower)
5 Conveyor bucket
6 Mixer (btwn. tower legs)
7 Bucket guides
11 Concrete chute
12 Receiving bin
13 Body cars

Figure 11.12 Construction of the bridge. (From Joyce, W.E., and M. Bebarfald, "Building the Rondout Creek Highway Suspension Bridge," *Engineering News Record*, 89, no. 11, 424–428, September 14, 1922.)

Figure 11.13 Continuous stiffening truss.

- The tower design is unique in that the legs are battered to provide clearance for the stiffening truss while also improving the transverse stability (see Figure 11.2). This was one of the first bridges to employ this feature, having been introduced into suspension bridge practice by H. D. Robinson for his proposed 1911 design of the Rhine Bridge in Cologne.
- The bridge employed construction techniques considered novel at the time, including the use of fixed saddles in lieu of roller nests. While fixed saddles provide for simpler and safer construction, they require the towers to be tipped back from the vertical in order to compensate for the elongation of the backstays (in this case, the tops of the towers were tipped back 6 in by means of temporary backstays). Another time-saving feature was the use of light false work to support the side spans during erection.

11.4 GENERAL INSPECTIONS

In the United States, highway bridge inspections and evaluations are mandated by the Federal Highway Administration (FHWA), as well as individual states, with procedures detailing inspection methods, organization responsibilities, inspector qualifications, and inspection frequency, reporting, and documentation, among others. The current National Bridge Inspection Standards (NBIS) detail the federal requirements for highway bridge inspections. The last revision to NBIS was done in 2004 and became effective in January 2005 [16]. NBIS set minimum standards for the inspection of all publicly owned highway bridges and is intended to ensure public safety by assuring that bridges have enough capacity to carry allowable loads. Many states and owners augment these data with more background needed for decisions taken to support their planning, design, maintenance, replacement, and rehabilitation activities. For example, New York State collects element-level inspection on a span basis [17].

New York State has some of the most rigorous inspection standards for highway bridges that are respected worldwide. The visual inspection requirements exceed those set by the FHWA requiring bridge inspections to be performed on all highway bridges on at least a biennial basis. All bridge inspection team leaders, quality control engineers, quality assurance engineers, and program managers are professional engineers with several years of bridge engineering–related experience.

The Wurts Street Bridge is inspected on an annual basis, rather than biennially, as it is weight restricted to vehicles weighing less than 3 U.S. tons. Essentially only cars can use the bridge in its present status. The inspections are hands-on due to its fracture critical status as well as fatigue-prone details. The special emphasis details include, but are not limited to, the suspension bridge span with stiffening trusses, floor beams, splay casting and anchor bars, welds on the bottom flange of stringers to the diagonal bracing channel, steel piers, and riveted connections. A hands-on inspection requires that bridge inspectors be at hand's reach and able to touch all these components during every inspection. All elements in each span receive an element condition rating on a scale of 7 (new) through 1 (failed). The overall condition rating of the bridge, a weighted rating of 13 main structural components, is calculated for all highway bridges in New York State.

NYSDOT also has a systematic procedure in place to identify both structural and nonstructural deficiencies that can affect public safety. The "flagging procedure" has been in place since 1985 and sets forth a uniform method of timely notification to responsible parties of serious bridge deficiencies that require immediate attention as well as issues that, if left unattended for an extended period, could become a serious problem in the future. For serious deficiencies, it further establishes requirements for certifying that appropriate corrective or protective measures are taken within a set time frame.

The critical inspection findings (flags) can be either structural or safety related. The structural flags are further subdivided into two categories: red structural flags and yellow structural flags. Red structural flags are used to report the failure of a critical primary structural component or a failure that is likely to occur before the next scheduled inspection. Yellow structural flags are used to report a potentially hazardous condition that, if left unattended beyond the next anticipated inspection, would likely become a clear and present danger. This flag is also used to report the actual or imminent failure of a noncritical structural component, where a failure may reduce the reserve capacity or redundancy of the bridge, but would not result in a structural collapse by the time of the next scheduled inspection interval. Nonstructural conditions are reported using a "safety flag" [17].

As noted above, Wurts Street Bridge is inspected on an annual basis as required by New York State inspection requirements, due to its condition and load restriction. Most recently, it has been inspected during October 2014 [18]. It took an inspection team a full week (approximately 70 h) to inspect the bridge. The overall (weighted) condition rating of this bridge is 3.641 based on this inspection indicating moderate deterioration of the bridge. Primary members, wearing surface, structural deck, paint, and bearings received a rating of 3, indicating deteriorated structural condition of the bridge and the need for 3-ton posting.

11.5 INTERNAL INSPECTIONS

Suspension bridge cables carry the weight of the deck and most of the imposed live load. As noted earlier, the suspension system is fracture critical and load path nonredundant. Hence, their condition assessment and evaluation is very critical in estimating the safety and remaining life of the structure. The cables are generally wrapped and cannot be visually inspected. Moreover, they are prone to deterioration due to ingress of moisture, hydrogen embrittlement, and other factors. At the same time, opening the cables and inspecting them completely is almost an impossible task due to time, cost, and effort required. Hence, opening small sections of representative cable sections, and evaluating them, while estimating the strength of the system by using statistical approaches has been used in recent years.

NCHRP Report 534 [19] provides guidelines for such cable inspection process, as including wire sampling and testing. This report provides complete instructions, illustrated with examples, using the condition and properties of the cable wire, determined by inspection and subsequent laboratory testing, for estimating cable strength. Based on the conditions observed, it also provides suggestions for frequency of such inspections.

In 1993, a limited internal inspection of a single panel of the main cable near midspan of the east side of the east cable was performed following a vehicle accident that resulted in damage to the wrapping wire [15]. In addition to this effort, the first major inspection of the main cables of the Wurts Street Bridge was performed in March and April of 2008 as part of a major evaluation of the bridge. Modjeski and Masters, Inc. (M&M) and contractor Piasecki Steel Construction, Inc., were retained by NYSDOT to conduct these inspections. M&M also documented all of these efforts and provided detailed reports to the NYSDOT.

11.5.1 1993 Internal inspections

The internal inspection [15] at the low point of the main-span east cable (PP1-2N) was conducted due to a vehicle impact that damaged the wrapping wire surrounding the cable. The repair involved removing

Figure 11.14 Cable condition from 1993 inspection. (Courtesy of William Moreau.)

the existing wrapping wire, thus providing an opportunity to examine the main cable wires, which are normally hidden under the protective wrapping. Once the wrapping wire was removed, the cable was wedged open using wooden wedges to facilitate visual inspection of the exposed wires.

The observed conditions (see Figure 11.14) showed black surfaces on the outer wires (indicating areas of depleted galvanizing) with only minimal areas of white powder (zinc oxide corrosion product) limited to the lower half of the cable. Most importantly, no ferrous corrosion or broken wires were observed. After the cable was inspected, the cable was oiled and then rewrapped. This panel was reopened as part of the 2008 inspection and the condition of the cable appeared to be nearly the same as it was 15 years earlier.

11.5.2 2008 Internal inspections

As noted above, the internal inspection of cables in 2008 [15] were performed by following NCHRP Report 534 guidelines. The locations were selected based on the recommendations contained in Report 534, which included looking for external signs of internal deterioration, such as loose wrapping wire, water dripping from the cable interior, rust stains, damaged caulking at the cable bands, surface ridges that indicate crossing wires beneath the wrapping wire, and a hollow sound emitted when the surface is tapped. Locations were also selected based on their placement along the cable. Historically, the worst cable deterioration has been observed near the quarter points of the cable, although damage has also been observed near low points. One of the locations selected was the same panel that was opened under the 1993 inspection (see Table 11.1).

Table 11.1 Panel locations for internal inspection

Cable	Span	Begin panel point	End panel point
East	North main span	1N	2N
	South main span	8S	9S
	South main span	16S	17S
	South side span	25S	26S
West	North side span	28N	29N
	North main span	14N	15N
	North main span	4N	5N
	South side span	28S	29S

Source: Modjeski and Masters, *The Wurts Street Bridge over Rondout Creek: 2008 Cable Strength Evaluation, Phase 2 Final Report*, submitted to New York State Department of Transportation, Poughkeepsie, New York, March 2011.

The cables were wedged at eight locations around the perimeter and inspected full length between the cable bands (see Figures 11.15 and 11.16). The internal inspection found the following general conditions to be similar on all the panels inspected [15]:

- The surface of the cable had a thick coating of red lead paste that was pliable and well adhered to the outer wires.
- Once the red lead had been removed to facilitate inspection, it was observed that the outer wires were dark in color, indicating a loss of the original galvanizing. However, no white oxidation product or ferrous corrosion was observed on the outer wires in the majority

Figure 11.15 Internal inspections during 2008. (From Modjeski and Masters, *The Wurts Street Bridge over Rondout Creek: 2008 Cable Strength Evaluation, Phase 2 Final Report*, submitted to New York State Department of Transportation, Poughkeepsie, New York, March 2011. With permission.)

Figure 11.16 Interior of cable during 2008 inspections showing grease from original construction. (From Modjeski and Masters, *The Wurts Street Bridge over Rondout Creek: 2008 Cable Strength Evaluation, Phase 2 Final Report*, submitted to New York State Department of Transportation, Poughkeepsie, New York, March 2011. With permission.)

of the locations. In the only two panels where these conditions were observed (panel points 1N–2N and 4N–5N), the zinc oxidation was found in localized patches only and the stage 3 wires had very minor spots of ferrous corrosion with no section loss.

- The interior of the cable was thoroughly coated with a thick grease known as petrolatum (similar in consistency to Vaseline but containing additives such as linseed oil), which was applied to the wires during the original aerial spinning of the cable (the original contract drawings refer to this as "slushing oil"). This material was used on other early suspension bridges (e.g., the Brooklyn Bridge) and is known to contain corrosion-inhibiting and water-displacing additives.
- Water was observed on the interior of the cable at several panel points. However, due to the protection afforded by the grease, only minor surface corrosion was found on a few interior wires.
- No broken wires were observed in any of the panels (see Table 11.2).

The cables were classified according to NCHRP criterion, samples were taken, and laboratory testing was performed by M&M. As noted in Table 11.2, the inspections found no wires in stage 4 (the worst stage, which implies that brown rust covering greater than 30% of the wire surface of a 3 to 6 in length of wire). See NCHRP Report 534 [19] for explanation of various stages.

The strength of the cable was determined following the NCHRP Report 534 guidelines. Table 11.3 provides cable strength at each location. This

Table 11.2 Cable condition summary

| Cable | Panel | Percent of total wire area | | | | Broken wires |
		Stage 1	Stage 2	Stage 3	Stage 4	
East	1N–2N	78.6%	17.7%	3.6%	0.0%	0
	8S–9S	92.6%	7.4%	0.0%	0.0%	0
	16S–17S	92.6%	7.4%	0.0%	0.0%	0
	25S–26S	94.6%	5.4%	0.0%	0.0%	0
West	28N–29N	92.9%	7.1%	0.0%	0.0%	0
	14N–15N	79.5%	20.5%	0.0%	0.0%	0
	4N–5N	91.5%	6.5%	2.0%	0.0%	0
	28S–29S	86.0%	14.0%	0.0%	0.0%	0

Source: Modjeski and Masters, *The Wurts Street Bridge over Rondout Creek: 2008 Cable Strength Evaluation, Phase 2 Final Report*, submitted to New York State Department of Transportation, Poughkeepsie, New York, March 2011.

Table 11.3 Cable strength summary

| Cable | Panel | Cable capacity[a] (kips) | | Percent loss |
		As built (1921)	Current (2008)	
East	1N–2N	11,928	11,908	0.2%
	8S–9S	11,928	11,928	0.0%
	16S–17S	11,928	11,928	0.0%
	25S–26S	12,847	12,847	0.0%
West	28N–29N	12,847	12,847	0.0%
	14N–15N	11,928	11,928	0.0%
	4N–5N	11,928	11,917	0.1%
	28S–29S	12,847	12,847	0.0%

Source: Modjeski and Masters, *The Wurts Street Bridge over Rondout Creek: 2008 Cable Strength Evaluation, Phase 2 Final Report*, submitted to New York State Department of Transportation, Poughkeepsie, New York, March 2011.

[a] Simplified strength model.

analysis indicated that the lowest cable capacity represents a reduction of only 0.2% from the as-built condition. The lowest factor of safety was found to be 3.90, which was close to the original as-built condition. These results indicated that cables have suffered no measurable loss of capacity in their more than 90 years in service. The report suggested that the cables will provide at least 50 more years of service based on observations, and suspenders are expected to last between 35 and 65 years.

11.5.3 Discussion

Some engineers [20] theorize that the relatively small-diameter main cable, not routinely caulked or painted, allowed any moisture that entered the cable to easily leak out. Without trapped moisture, many oxidation and chemical reactions just do not occur to degrade the main cable.

A second theory, which has gained some lab-based confirmation, is that the stress level in the main cable wires is a contributing factor in the rate of hydrogen-assisted degradation and/or loss of ductility. The Wurts Street Bridge was designed with main cables having a safety factor of nearly 4. Modern bridges, post 1975, typically are designed with safety factors of 2.25 or less in the main cable.

A number of factors may also be contributing to the health of the main cables at the Wurts Street Bridge; these cables are now over 90 years old, while many suspension bridges only half that age are showing advanced signs of main cable degradation. When main cable conditions are compared to those of suspension bridges of comparable age and location, the Wurts Street Bridge (built in 1921) cables are in far better shape than the Mid-Hudson (built in 1930) and Bear Mountain (built in 1924) Bridges.

The Wurts Street and Bear Mountain Bridges' cables were manufactured by Roebling and the Mid-Hudson Bridge's cables were manufactured and spun by American Bridge. Correlating the aging effects on these main cables is challenging with so many interrelated cause-and-effect relationships. History has not confirmed any of the simple relationships in rates and extent of main cable degradation between similar bridges.

A thorough evaluation of any main cable condition assessment must take into account all site-specific corrosion amplifiers, microclimates, and stress risers specific to the design and construction of the bridge at hand.

11.6 PAST WORK

The bridge was rehabilitated in 1974. This work included painting, steel repairs, replacement of concrete deck, and railing replacement. The structural steel was painted in 1990. No major work was done on the bridge except for routine maintenance for the last 24 years. The maintenance work includes items such as bridge cleaning, railing repair, joint repair, touch-up of structural paint, and repair of deteriorated connections.

11.7 SUMMARY

This chapter provides history, design and construction details, and current conditions of the Kingston–Port Ewen Bridge (Wurts Street Bridge), built in 1921 and located in the mid-Hudson region of New York State. Although it may be a relatively small suspension span, this structure demonstrated the adaptability and economic viability of suspension bridge design and construction principles to relatively short-span highway bridges in the early 20th century. The bridge has survived for more than 90 years and still serves the region well, showcasing one of the historical landmarks for infrastructure achievements of modern civil engineering.

ACKNOWLEDGMENTS

All of the opinions and findings reported in this chapter are those of the author and not of the organization that the author represents. The author acknowledges Jane Minnotti for her assistance with literature searches. Richard Marchione, David Bennett, Eric Foster, Francois Ghanem, Ellen Zinni, and Rodney Delisle of NYSDOT reviewed the chapter and provided comments.

REFERENCES

1. http://en.wikipedia.org/wiki/Rondout_Creek_bridge (accessed on November 29, 2014).
2. http://cr.nps.gov/nr/travel/kingston/k22.htm (accessed on November 29, 2014).
3. "Big Bridge for Kingston," *New York Times*, October 17, 1915.
4. "Heavy Motor Traffic Needs Better Roads," *New York Times*, January 25, 1920.
5. "Rondout Bridge to Open," *New York Times*, November 15, 1921.
6. "Rondout Bridge Dedicated," *New York Times*, November 30, 1921.
7. "New Highway Bridge," *New York Times*, December 4, 1921.
8. "The Rondout Bridge," *New York Times*, December 6, 1921.
9. "Rondout Bridge Open," *New York Times*, May 14, 1922.
10. "Rondout Creek Bridge Design Condemned for Excessive Load on Batter Piles," *Engineering News Record*, Vol. 88, No. 7, pp. 329–332, August 14, 1919.
11. Joyce, W.E., and Bebarfald, M. "Building the Rondout Creek Highway Suspension Bridge," *Engineering News Record*, Vol. 89, No. 11, pp. 424–428, September 14, 1922.
12. Dave Bennett, "Wurts Street Over Roundout Creek" (Personal Communications).
13. "Weight and Cost of Rondout Creek Bridge," *Engineering News Record*, Vol. 89, No. 21, p. 874, 1922.
14. Hool, A.G., and Kinne, W.S. *Movable and Long-Span Steel Bridges*, revised by Zippoldt, R.R., and Langley, H.E., McGraw-Hill Book Company, Inc., New York, 1943.
15. Modjeski and Masters. The Wurts Street Bridge over Rondout Creek: 2008 Cable Strength Evaluation, Phase 2 Final Report, submitted to New York State Department of Transportation, Poughkeepsie, New York, March 2011.
16. National Bridge Inspection Standards, 23 CFR Part 650. Federal Highway Administration, U.S. Department of Transportation, Federal Register, Vol. 69, No. 239, 2004.
17. Bridge Inspection Manual, New York State Department of Transportation, Albany, New York, 2014.

18. 2014 Biennial Inspection Report for BIN 1007350, New York State Department of Transportation, Albany, New York, 2014.
19. Guidelines for Inspection and Strength Evaluation of Suspension Bridge Parallel Wire Cables, Report 534, National Cooperative Highway Research Program, Washington, DC, 2004.
20. Moreau, W.J. (Personal Communications), March 2015.

Index

Page numbers followed by f and t indicate figures and tables, respectively.